ALGEBRA

Part II. Expressions

Lessons for Self-Study with Test Preparation

Build Your Self-Confidence and Enjoyment of Math!

All about Expressions in Algebra

with a comprehensive Solutions Manual

Aejeong Kang

MathRadar

Send all inquiries to:

MathRadar, LLC
5705 Spring Hill Dr.
Mckinney, Texas 75072

Visit www.mathradar.com for more information and a sneak preview of the MathRadar series of math books.

Send inquires via email at info@mathradar.com

Algebra, Part II: Expressions

ISBN-13: 978-0-9893689-1-9

ISBN-10: 0989368912

Printed in the United States of America.

Preface

I wrote these books because I am a mother and I have a strong academic background in mathematics. I have a BS degree in Mathematics and Master's degree in Mathematics as well. I have completed Ph.D. program in Biostatistics.

After receiving the big blessing of our first child, a daughter, I decided to forgo my personal career goals to become a full-time mother. When our daughter entered 7th grade, that meant lots of help with her study of math-my passion. However, I struggled to find good math books that would help her understand difficult concepts both clearly and quickly. After the conversation with my husband and (now two) children, I decided that the best way to help my children was by writing math books for them myself. They wholeheartedly agreed.

That's why I've been able to pour all my knowledge, energy, and soul into these books. Because I'm a mom, I would do anything for my children. Thanks to my family's endless support, I wrote them four books, designed for use in junior high and high-school (partially) mathematics.

And that would have been the end of my journey, but my husband and children insisted that I share my work outside of our family. They encouraged me to make my work available to other parents looking, as I was, for well-written, great mathematics books for their children.

So I finally decided to publish these books. I do so with the hope that they will help your children find success and confidence in learning and studying mathematics.

But I would never have begun or finished this project without the support of my family. Kyungwan, Nichole, and Richard, you are my world.

Thank you.

Introduction

After reading several pages of explanation/description about a certain mathematical concept, you still don't get it.

You have worked on many related problems to understand mathematical concepts, but you still feel completely lost in the mathematical jungle.

You bought a math book with good reviews, but it only offers short answers without detailed solutions. You feel confused and frustrated.

You've tried multiple learning math books, but you've still not getting good grades in math. It seems like math is just not for you.

If any one of these situation sound familiar, the MathRadar series will help you escape!

Everyone has different learning abilities and academic skill. The MathRadar series is written and organized with emphasis on helping each individual study mathematics at his/her own pace.

In the case of Algebra, each book covers all the topics required in each field of Algebra. These fields were systematically subdivided into Part I, Part II, and Part III.

Algebra, Part I covers information about **Number Systems** from natural numbers to real numbers.

> **Level I**   for **grades 6~8** : Chapter 1 and Chapter 2
>
> **Level II**   for **grades 7~9** : Chapter 3
>
> **Level III** for **grades 8~10** : Chapter 4

Algebra, Part II covers information about **Expressions** dealing with equations and inequalities.

> **Level I**   for **grades 6~8** : Chapter 1 and Chapter 2
>
> **Level II**   for **grades 7~9** : Chapter 3, Chapter 4, and Chapter 5
>
> **Level III** for **grades 8~10** : Chapter 6, Chapter 7, and Chapter 8

Algebra, Part III covers information about **<u>Functions</u>**, while also including **<u>Statistics and Probability</u>**.

Level I	for **grades 6~8** : Chapter 1	**Statistics** : Chapter 1, Chapter 2, and Chapter 3	
Level II	for **grades 7~9** : Chapter 2	**Probability** : Chapter 4	
Level III	for **grades 8~10** : Chapter 3		

Each book consists of clean and concise summaries, callouts, additional supporting explanations, quick reminders and/or shortcuts to facilitate better understanding.

With the numerous examples and exercises, students can check their comprehension levels with both basic and more advanced problems.

Each book includes **<u>Solutions Manual.</u>** The solutions manual makes it possible for students to study difficult concepts on their own. With the solutions manual, students will be able to better understand how to solve problems through step-by-step for each problem.

Geometry has also been systematically subdivided so that students can easily grasp geometry concepts. Each concept is thoroughly explained with step-by-step instruction and detailed proofs.

Carry the MathRadar series with you!

Work on them anytime and anywhere!

Finally, you can start to enjoy mathematics!

Whether you are struggling or advanced in your math skills, the MathRadar series books will build your self-confidence and enjoyment of math.

I hope Math Radar is what you need and will be a great tool for your hard work.

Your comments or suggestions are greatly appreciated.

Please visit my website at www. mathradar.com or email me at aejeong@mathradar.com

Thank you very much. And remember, math can be fun!

Aejeong Kang

How to use the MathRadar series

Once you finish level I of the MathRadar Algebra book of your choice, you will have 2 options to choose from.

Option 1. Keep working on level II and III of the same book
to complete your study up to the beginning of high school level.

Option 2. Start working on level I in other MathRadar books
to complete your current level in other areas of Algebra.

Since the MathRadar Series is written to meet with everyone's different learning abilities and academic skills, and not for the grade level or age, you can build your progress quickly with strong confidence. Once you complete all the levels of the MathRadar series, you can see yourself on highly advanced level of mathematics. Then you can continue to learn anything very easily with a strong foundation and base.

The MathRadar series of math books will be the beneficial and valuable resources towards your goal.

Level I : Grades 6-8

Level II : Grades 7-9

Level III : Grades 8-10

➡ EXPRESSIONS

TABLE OF CONTENTS

Chapter 3   Monomials and Polynomials

level II

Chapter 4   Systems of Equations

level II

Chapter 5   Systems of Inequalities

level II

Chapter 6   Factorization

level III

Chapter 7   Quadratic Equations

level III

level III

Chapter 8   Rational Expressions (Algebraic Functions)

Solutions Manual

Index

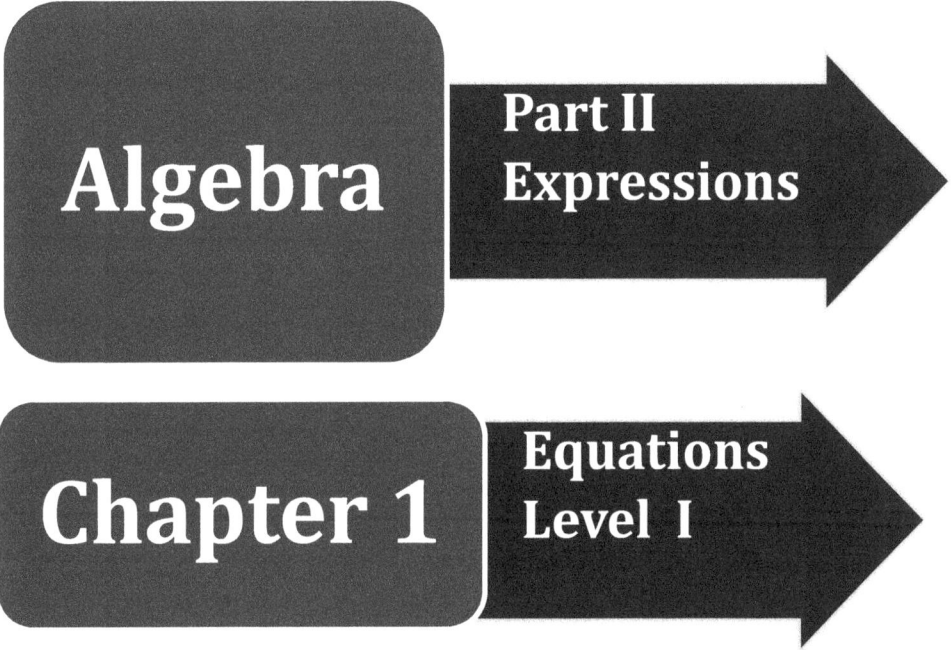

Algebra

Part II Expressions

Chapter 1

Equations Level I

CHAPTER 1

Chapter 1   Equations

1-1 Variables and Expressions

  1.   Definition

1-2 Equations

1. Definition

2. Properties of Equations

3. Linear Equations and their Solutions

   (1) Linear Equations

   (2) Solutions

 4. Solving Linear Equations

   (1) Steps for Solving Equations with One Variable

   (2) Steps for Solving Word Problems

5. Absolute Values

6. Linear Equations with Absolute Values

Chapter1. Equations

1-1   Variables and Expressions

1. Definition

(1) Variable

A *variable* is a letter that represents an unknown number.

For example, $2x$ is the product of the number 2 and an unknown number named x.

(2) Expression

An *expression* is a number phrase without an equal sign or inequality sign.

(3) Term

A *term* is an expression of a single number or a product of numbers and variables.

For example, $3x + 5$ has two terms $3x$ and 5

(4) Constant

A *constant* is a term that doesn't contain any variable.

For example, 6 is the constant of the expression $4x + 5y + 6$

(5) Coefficient

A *coefficient* is a number that is multiplied by a variable

For example, the coefficient of $8x$ is 8.

(6) Monomial expression

A *monomial expression* is an expression of only one term which is product of numbers and variables.

For example, $2,\ x,\ 3x,\ \frac{1}{2}xy$

(7) Polynomial Expression

A *polynomial expression* is an expression of two or more terms combined by addition and/or subtraction.

For example, $3x + 4,\ \frac{1}{3}xy + 2x - 5$

1-2 Equations

1. Definition

An equation has two expressions separated by an equal sign (=). The value of the expression on the left side of the equal sign must be the same as the value of the expression on the right side.

An expression using an equal sign between two expressions is called an *equation*.

For example, $5 + 6 = 11$ and $2x + 4 = 10$ are equations. $3 + 4 = 8$ is also an equation.

But $2x + 4$ is not an equation because the expression doesn't have an equal sign.

true equation

depends on the value of x

false equation

2. Properties of Equations

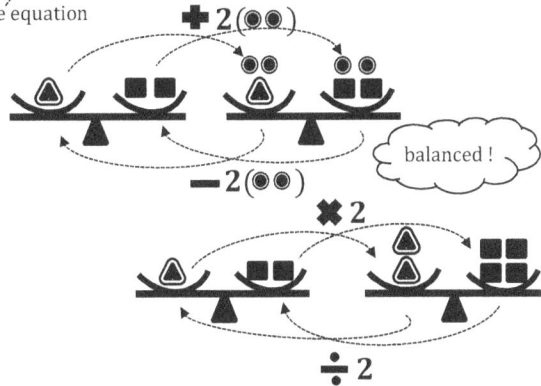

(1) For any c, $a = b \implies a + c = b + c$

$$a - c = b - c$$

$$a \times c = b \times c$$

$$a \div c = b \div c, \quad c \neq 0$$

balanced !

The expressions on both sides of the equal sign must have the same value.

(This keeps the equation balanced.)

$$a = b \Rightarrow ac = bc$$

$$ac = bc \not\Rightarrow a = b$$

$$a = b \Rightarrow \frac{a}{c} = \frac{b}{c}, c \neq 0$$

$$a = b \not\Rightarrow \frac{a}{c} = \frac{b}{c}$$

$$a \div b \times c = a \times \frac{1}{b} \times c = \frac{ac}{b}$$

$$a \div b \times c \neq a \div bc = \frac{a}{bc}$$

$$a : b = c : d$$

inside

outside

(2) $\frac{a}{b} = \frac{c}{d}$ ($a : b = c : d$) $\implies$ 1) $ad = bc$ ($\because$ cross product)

2) $\frac{a+b}{b} = \frac{c+d}{d}$ ($\because \frac{a}{b} = \frac{c}{d} \Rightarrow \frac{a}{b} + 1 = \frac{c}{d} + 1 \Rightarrow \frac{a+b}{b} = \frac{c+d}{d}$)

3) $\frac{a-b}{b} = \frac{c-d}{d}$ ($\because \frac{a}{b} = \frac{c}{d} \Rightarrow \frac{a}{b} - 1 = \frac{c}{d} - 1 \Rightarrow \frac{a-b}{b} = \frac{c-d}{d}$)

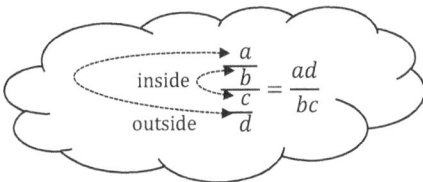

inside

outside

$$\frac{\frac{a}{b}}{\frac{c}{d}} = \frac{ad}{bc}$$

4) $\frac{a+b}{a-b} = \frac{c+d}{c-d}$ ($\because \frac{\frac{a+b}{b}}{\frac{a-b}{b}} = \frac{\frac{c+d}{d}}{\frac{c-d}{d}} \Rightarrow \frac{a+b}{a-b} = \frac{c+d}{c-d}$)

5) $\frac{a}{b} = \frac{c}{d} = \frac{e}{f} = \frac{a+c+e}{b+d+f}$, $b + d + f \neq 0$

$\because$ = because

$\therefore$ = therefore

($\because$ let $\frac{a}{b} = \frac{c}{d} = \frac{e}{f} = k$. Then $a = bk$, $c = dk$, $e = fk$.

$\therefore a + c + e = (b + d + f)k \quad \therefore k = \frac{a+c+e}{b+d+f}$, $b + d + f \neq 0$)

3. Linear Equations and their Solutions

(1) Linear Equations

A *linear equation* is an equation with one variable which has the highest power of 1.

For example, the equation $ax = b$, where a and b are constants, is a linear equation with a variable x.

(2) Solutions

The *solution* (*root*) of an equation is a number which makes the equation a true expression.

For example, the linear equation $2x = 10$ is true when $x = 5$.

Thus, $x = 5$ is the solution of the equation $2x = 10$.

For the linear equation $ax = b$, where a and b are constants, the solution is

1) $a \neq 0 \Rightarrow x = \dfrac{b}{a}$ (Only one solution)

2) $a = 0 \Rightarrow$ ① $b \neq 0 \Rightarrow$ No solution ($\because$ Division is not defined.)

② $b = 0 \Rightarrow$ All real numbers

4. Solving Linear Equations

For $ax = b$,

$0 \cdot x = 0 \implies x = \dfrac{0}{0}$ is the unlimited solution.

$0 \cdot x = b\ (\neq 0) \implies x = \dfrac{b}{0}$ is not defined. There is no solution.

(1) Steps for Solving Equations with One Variable

Step 1 : If the coefficient of the equation is a fraction or decimal, change it to an integer by multiplying a proper number by both sides of the equation.

Step 2 : If there are parentheses, remove them by using the distributive property.

Step 3 : Convert it into a simpler equation; transfer all variables to one side of the equation and transfer all numbers to the other side of the equation in the form of $ax = b$, $a \neq 0$.

Step 4 : Solve for the variable x by dividing each side of the equation ($ax = b$) by a.

This gives $x = \dfrac{b}{a}$, $a \neq 0$. Therefore, $\dfrac{b}{a}$ is the one and only solution of $ax = b$.

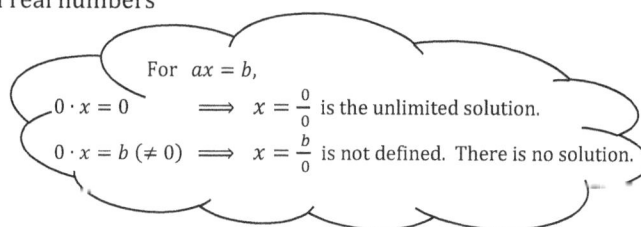

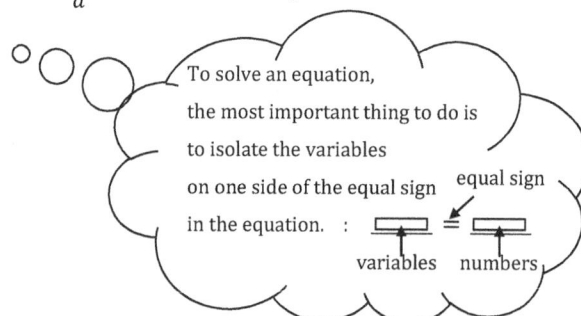

To solve an equation, the most important thing to do is to isolate the variables on one side of the equal sign in the equation. :

equal sign

variables numbers

Example 1

$$2x + 3 = 7 \xRightarrow[\text{Subtract 3 from both sides}]{} 2x + 3 - 3 = 7 - 3$$

$$\xRightarrow[\text{variable=number}]{} 2x = 4$$

$$\xRightarrow[\text{Multiply each side by } \frac{1}{2} \text{ or divide each side by 2}]{} 2x \times \frac{1}{2} = 4 \times \frac{1}{2} \text{ or } \frac{2x}{2} = \frac{4}{2}$$

$$\xRightarrow[\text{reduce to lowest terms}]{} x = 2$$

Example 2

$$\frac{5}{2}x + 4 = x - 2 \xRightarrow[\text{Multiply each side by 2}]{} 5x + 8 = 2x - 4$$

$$\xRightarrow[\text{transfer (Step 3)}]{} 5x - 2x = -4 - 8$$

$$\xRightarrow[\text{simplify}]{} 3x = -12$$

$$\xRightarrow[\text{Divide each side by 3 or multiply each side by } \frac{1}{3}]{} \frac{3x}{3} = \frac{-12}{3} \text{ or } 3x \times \frac{1}{3} = -12 \times \frac{1}{3}$$

$$\xRightarrow[\text{simplify}]{} x = -4$$

(2) Steps for Solving Word Problems

1) Assign a variable to represent the unknown number.

2) Find an equation for the problem.

3) Solve the equation.

4) Check the solution by substituting the solution into the variable in the equation.

Solution Problems:

The salt refers to as the solute, the water is the solvent, and the resulting mixture as the solution (water+salt). The amount (the concentration) of salt in the solution expresses how salty the salt water is.

$$\text{Concentration} = \frac{\textbf{The amount of salt (solute)}}{\textbf{The total amount of solution}}$$

Concentration is normally expressed as a percent (%), multiplied by 100.

The amount of salt (solute)
= (**Concentration**) × (**The total amount of solution**)

Distance, Rate, and Time Problems:

Distance = Rate × Time

$$\text{Rate} = \frac{\text{Distance}}{\text{Time}}, \ \text{Time} = \frac{\text{Distance}}{\text{Rate}}$$

If the rate is in miles per hour, then the distance must be in miles and the time in hours.
If the time is in minutes, convert it to hours (dividing by 60) to find the distance in miles.

Match the units!!

---

5. Absolute Values

The absolute value of a number a is denoted by $|a|$.

If $a \geq 0$, then $|a| = a$

If $a < 0$, then $|a|$ is the corresponding positive number.

The absolute value of a number a, denoted by $|a|$ is the distance of a from zero on a number line. Since a distance is always a positive number, absolute value is never negative.
For example,
$|2| = 2$
: two units to the right of zero on a number line.
$|-2| = 2$
: two units to the left of zero on a number line.

Example

$|0| = 0, \ |1| = 1, \ |2| = 2, \ |-1| = 1, \ |-2| = 2$

If the number has no minus sign in front of it, the absolute value is not changed.

But if the number has minus sign in front of it, we remove the minus sign to find the absolute

Value. For example,

if $a = -1$, then $-a = -(-1) = 1$ and if $a = -2$, then $-a = -(-2) = 2$.

Therefore, we define the absolute value of a as

$$|a| = \begin{cases} a & \text{if } a \geq 0 \\ -a & \text{if } a < 0 \end{cases}$$

$|-2| = -(-2) = 2$
$-|-2| = -2$

6. Linear Equations with Absolute Values

The absolute value of a real number a, denoted by $|a|$, is

$$|a| = \begin{cases} a & \text{if } a \geq 0 \\ -a & \text{if } a < 0 \end{cases}$$

Example 1

$|x - 1| = 2x + 3$

$$\implies \begin{cases} x - 1 \geq 0 \ (x \geq 1) \ \Rightarrow \ |x-1| = x - 1 = 2x + 3 \ ; x = -4 \ ; \text{not possible} \\ x - 1 < 0 \ (x < 1) \ \Rightarrow \ |x-1| = -(x-1) = 2x + 3 \ ; x = -\frac{2}{3} \end{cases}$$

$\therefore \ x = -\frac{2}{3}$

Example 2

$|x + 1| + |x + 2| = 5$

Since $(x + 1 = 0 \Rightarrow x = -1)$ and $(x + 2 = 0 \Rightarrow x = -2)$,

Consider $x < -2, \ -2 \leq x < -1, \ x \geq -1$

$$\implies \begin{cases} \text{①} \ x < -2 \ ; \ -(x+1) - (x+2) = 5 \ ; \ x = -4 \\ \text{②} \ -2 \leq x < -1 \ ; -(x+1) + (x+2) = 5 \ ; \ 0 \cdot x = 4 \ ; \text{no solution} \\ \text{③} \ x \geq -1 \ ; \ x + 1 + x + 2 = 5 \ ; x = 1 \end{cases}$$

$\therefore x = -4, \ x = 1$

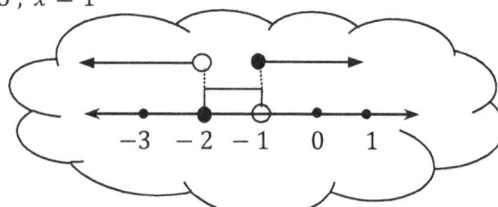

Example 3

$|x + 1| = |x + 3|$

$\implies x + 1 = \pm (x + 3)$

$\circ \circ \circ$ $|a| = |b| \Rightarrow a = \pm b$

$$\implies \begin{cases} x + 1 = x + 3 \implies 0 \cdot x = 2 \ ; \text{no solution} \\ x + 1 = -(x + 3) \implies 2x = -4 \ ; x = -2 \end{cases}$$

$\therefore \ x = -2$

Exercises

 # 1. Find the value for each expression.

(1) $x + 3$ if $x = -3$

(2) $-(x + 1)$ if $x = -5$

(3) $x^3 - 2x - 5$ if $x = -1$

(4) $2xy - 4$ if $x = -3$, $y = 2$

(5) $x^2 - 5y$ if $x = -2$, $y = -3$

(6) $\frac{2}{x} + \frac{3}{y}$ if $x = -\frac{1}{4}, y = -\frac{1}{6}$

(7) $\frac{y}{x} - \frac{x}{y}$ if $x = 2$, $y = -3$

(8) $x^{99} - x^6$ if $x = -1$

(9) $\frac{3}{a} - 2b^2$ if $a = \frac{1}{4}$, $b = -3$

(10) $\frac{1}{a} - \frac{2}{b} - \frac{3}{c}$ if $a = -\frac{1}{4}$, $b = -\frac{1}{2}$, $c = -6$

(11) $2(a - b) - (a^2 - b^2)$ if $a = -2$, $b = 3$

(12) $|a - 2| - |3ab - a|$ if $a = -1$, $b = 2$

 # 2. Simplify each expression.

(1) $\frac{1}{2}x \cdot (-6)$

(2) $-\frac{2}{3}(6x - 9)$

(3) $(6x - 2) \div \left(-\frac{3}{10}\right)$

(4) $\frac{1}{2}(4x - 6) - \frac{2}{3}\left(9x - \frac{3}{4}\right)$

(5) $-3x + 6x - 2x - 5$

(6) $2x + 4y - \{3x - (5 - 2y)\} - 3$

(7) $(2a - 3b) - (5a - 2b) - (4 - a)$

(8) $3m^2 - (5m - m^2 - 1) + 2m$

(9) $\frac{3t^3 - 4t^2}{2t}$

(10) $\frac{2a^2b - 3ab^2 - 5ab}{ab}$

(11) $(3x - 9) \div \frac{3}{2} - 8\left(\frac{3}{4}x - 2\right)$

(12) $\frac{x-2}{3} - \frac{2x-1}{4} - \frac{3-x}{2}$

 # 3. Find an expression for the perimeter.

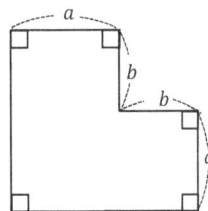

4. Find an expression for the shaded area.

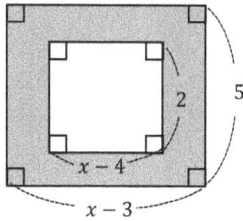

5. Which expression is different from the others?

(1) $x \div (y \times z)$

(2) $x \div y \div z$

(3) $x \times \dfrac{1}{y} \div z$

(4) $x \div (y \div z)$

(5) $x \times \dfrac{1}{y} \times \dfrac{1}{z}$

6. Find $a + b$ and $a - b$

when a is a coefficient of x, and b is a constant for

the expression $\dfrac{4x-3}{2} - \dfrac{2x-1}{3}$

7. The coefficient of x is 3 and the constant is 5

for the form $2x + b - (ax + 3)$. Find $a \cdot b$ and $\dfrac{a}{b}$

8. For two expressions A and B,

if you add $2x + 3$ to A then you get $5x + 7$ and

if you subtract $-3x - 4$ from B then you get $2x + 5$. What is $2A - 3B$?

9. Solve the following equations for x :

(1) $3x - 2 = 7$

(2) $2x + 3 = 3x - 2$

(3) $5x - 2 = \dfrac{1}{2}x - 1\dfrac{1}{4}$

(4) $0.2x - 0.3 = 0.4x - 0.5$

(5) $\dfrac{3}{4}\left(x - \dfrac{1}{3}\right) = \dfrac{1}{2}\left(\dfrac{1}{5} + 4x\right)$

(6) $\dfrac{x-3}{2} - 1 = \dfrac{x}{4} - 3$

(7) $3(1 - 2x) + 7 = -2x - 2$

(8) $3 - \dfrac{2x-1}{3} = 5x - \dfrac{x-2}{6}$

\# 10. Find a constant a that makes the equation $4(a + x) = 2(2x + 3) + 6$ true for all values of x.

\# 11. For any positive integers $a, b,$ and c, a is divided by b and the remainder is c. Express the quotient using $a, b,$ and c.

\# 12. For any non-zero constants a, $b(b \neq 1)$, and c, find the value of $\dfrac{1}{abc}$ such that $a + \dfrac{1}{b} = 1$ and $b + \dfrac{1}{c} = 1$.

\# 13. $a = \dfrac{2}{3}, b = \dfrac{3}{4},$ and $c = -\dfrac{4}{5}$.

Find the value of $\dfrac{ab+bc+ca}{abc}$.

\# 14. Find the sum of all possible solutions for the equation $|2x - 3| = 5$.

\# 15. $A = 2x - 3$, $B = 3x + 4$.

The ratio of A to B is $3 : 5$. When a is the solution of x, find the value of $-\dfrac{a}{3} + 3$.

\# 16. For any constants a, b, the solution of the equation $3x - 2 = ax - 4$ is $x = -1$ and the solution of the equation $\dfrac{1}{2}x + b = ax + 3$ is $x = -2$. Find $a \cdot b$

\# 17. The solution of the equation $2ax + 5 = -3$ is half of the solution of the equation $x - 5 = 3x + 7$. Find the value of $3a - 4$.

\# 18. For any constants a and b, $\frac{1}{a} - \frac{1}{b} = 3$ $(ab \neq 0)$. Find the value of $\frac{5a - 3ab - 5b}{a - b}$.

\# 19. $\begin{cases} (1) \ \frac{a+3}{4} - \frac{2x-2}{3} = 1 \\ (2) \ \frac{3a-2}{2} - \frac{2a-x}{3} = 1 \end{cases}$

When the ratio of the solution of (1) to the solution of (2) is $1 : 4$, find the value of a.

\# 20. For any x, the equation $3x - 5a = 2bx + 6$, where a and b are constants, is always true. Find the value of $\frac{a}{2b}$.

\# 21. The solution of an equation $\frac{2x-5a}{3} + x + 4 = 8$ is a negative integer. Find the greatest value of a.

\# 22. $a@b = ab^2 + a^2 b$

When $\frac{1}{a} = 2$, $\frac{1}{b} = -3$, find the value of $b@a$.

\# 23. How much water should be added to 30 ounces of a 20% salt solution to produce a 15% solution?

\# 24. Richard has 20 ounces of a 15% of salt solution. How much salt should he add to make it a 20% solution?

\# 25. Richard drives to place A at 30 miles per hour. 20 minutes after he departs, Nichole goes to the place A at 50 miles per hour. How long will it take until Richard meets Nichole?

26. Richard wants to make 50 ounces of a 10% salt solution by mixing a 7% salt solution with a 15% salt solution. How many ounces of a 7% salt solution must be mixed?

27. Richard spends two-thirds of the money in his pocket to buy a book. He now has 4 dollars left. How much money did he have at the beginning?

28. A bag is on sale for a 15% discount. Nichole paid $60, including a 6% sales tax. What was the original price of the bag (rounded to the nearest hundredth)?

29. Richard's aunt is 51 years old. She is three times as old as the sum of the ages of Richard and his sister. Richard is 7 years younger than his sister. How old is Richard's sister?

30. The sum of three consecutive odd integers is 153. Find the biggest number of these three integers.

31. The tens digit of a certain two-digit integer is 3. If the digits of the number are interchanged, the number will be 1 less than two times the original number. Find the original number.

32 Richard is 5 years old and Nichole is 12 years old. In how many years will Nichole be two times Richard's age?

33. Richard takes 3 hours to finish a job if he works alone. Nichole takes 2 hours to finish the same job if she works alone. How long will it take them to finish the job if they work together?

34. Nichole took 8 days to finish a job and Richard took 6 days to finish the same job. If Nichole worked $3\frac{1}{3}$ days alone and then Nichole and Richard worked together to finish the job, how many days did they work together?

35. Nichole checked a book out from a library. She read $\frac{1}{3}$ of the book on the first day, $\frac{1}{4}$ of the book on the second day, and 39 pages on the third day. She now has to read $\frac{1}{5}$ of the book to finish. How many pages does the book have?

36. Richard finishes a job alone in 5 hours. If Nichole helps him, they can finish the job together in 1 hour 40 minutes. How many hours would it take Nichole to work alone to finish the job?

37. Nichole goes out to eat at a restaurant. Her total bill is $23, including a 15% tip. How much was the dinner?

\# 38. A movie ticket price for children is \$3 less than the adult ticket price. Nichole paid \$36 for 2 adults and 3 children. What is the price of an adult ticket?

\# 39. Nichole and Richard live in the same home. They drove to a park to meet some friends. They started from their home at the same time. Nichole drove at 40 miles per hour and Richard drove at 50 miles per hour. Nichole arrived at the park 10 minutes late while Richard arrived 5 minutes early for their appointment. Find the distance from Richard and Nichole's home to the park.

\# 40. Find the value for each of the following

(1) $|4|$

(2) $|-5|$

(3) $-|-3|$

(4) $|2 - 6|$

(5) $|-7 - 5|$

(6) $|5| + |-5|$

(7) $|3| - |8|$

(8) $|-9| + (-9)$

\# 41. Solve the following equations

(1) $|x - 3| = 5x + 2$

(2) $|x - 4| + |x + 2| = 10$

(3) $|x - 2| - |5 - x| = 0$

Algebra

**Part II
Expressions**

Chapter 2

**Inequalities
Level I**

CHAPTER 2

Chapter 2   Inequalities

Chapter 2. Inequalities

2-1   Inequalities with One Variable

1.   Algebraic Inequality Symbols

$<$: Less than

$>$: Greater than

$\leq$: Less than or equal to

$\geq$: Greater than or equal to

$\neq$: Not equal to

$a < b$ (a is less than b) $\Leftrightarrow b - a$ is positive

$a > b$ (a is greater than b) $\Leftrightarrow b - a$ is negative

$a \leq b \Leftrightarrow a < b$ or $a = b$

So $1 \leq 1$ is true .

But $1 < 1$ is false.

$a \leq a$

2.   Definition

Inequality is an expression using the algebraic inequality symbols to show the relationship between the values of numbers or variables.

Examples

(1) $x > 2$ (x is greater than 2) means

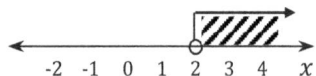

2 is not included (open circle) in the solution.

(2) $-2 \leq x < 1$ (x is greater than or equal to -2 and less than 1) means

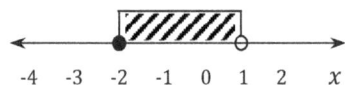

-2 is included (closed circle) in the solution.

1 is not included (open circle) in the solution.

3. Properties of Inequalities

For any real numbers a, b and c, the following properties apply

(1) Transitive Property

$$a < b, \ b < c \ \Rightarrow \ a < c$$

(2) Adding or subtracting the same number to or from each side of an inequality does not change the direction of the inequality symbol.

$$a < b \ \Rightarrow \ a + c < b + c \ \text{ and } \ a - c < b - c$$

(3) Multiplying or dividing each side of an inequality by the same positive number does not change the direction of the inequality symbol.

$$a < b, \ c > 0 \ \Rightarrow \ a \cdot c < b \cdot c \ \text{ and } \ \frac{a}{c} < \frac{b}{c}$$

$a < b, \ c = 0 \ \Rightarrow \ a \cdot c = b \cdot c$

$a < b \ \not\Rightarrow \ a^2 < b^2 \ (\because -2 < -1, \text{ but } (-2)^2 > (-1)^2)$

$0 < a < b \ \Rightarrow \ a^2 < b^2$

$a < b \ \Rightarrow \ a^3 < b^3 \ (\because -3 < -2, \text{ and } (-3)^3 < (-2)^3)$

(4) Multiplying or dividing both sides of an inequality by the same negative number will change the direction by reversing the inequality symbol).

$$a < b, \ c < 0 \ \Rightarrow \ a \cdot c > b \cdot c \ \text{ and } \ \frac{a}{c} > \frac{b}{c}$$

$-2x < 1 \xrightarrow[\times\left(-\frac{1}{2}\right)]{} x > -\frac{1}{2}$

$a < x \le b \xrightarrow[\times(-1)]{} -a > -x \ge -b \ ; \ -b \le -x < -a$

(5) $$a < b \iff a - b < 0$$
$$a > b \iff a - b > 0$$

(6) If a and b have the same sign $\Rightarrow a \cdot b > 0, \ \frac{a}{b} > 0, \text{ and } \frac{b}{a} > 0$

If a and b have different signs $\Rightarrow a \cdot b < 0, \ \frac{a}{b} < 0, \text{ and } \frac{b}{a} < 0$

$a < b \iff -a > -b$

(7) Expanded Properties

$$a < x < b \ \Rightarrow \ a + c < x + c < b + c, \ \ a - c < x - c < b - c$$

$$a < x < b, \ c > 0 \ \Rightarrow \ ac < xc < bc, \ \ \frac{a}{c} < \frac{x}{c} < \frac{b}{c}$$

$$a < x < b, \ c < 0 \ \Rightarrow \ ac > xc > bc, \ \ \frac{a}{c} > \frac{x}{c} > \frac{b}{c}$$

4. Solutions of Linear Inequalities

To solve a linear inequality with one variable (to identify the values of x which satisfy the inequality), isolate the variable on one side of the inequality and solve it exactly like a linear equation. The solution will consist of intervals or unions of intervals.

(1) The linear Inequality is formed by

$ax < b \, (\, ax \leq b\,)$ or $ax > b \, (\, ax \geq b\,)$ for any $a \neq 0$

(2) The solution of the inequality $ax < b$ is

① $x < \dfrac{b}{a}$, when $a > 0$ (positive a)

② $x > \dfrac{b}{a}$, when $a < 0$ (negative a)

③ all real numbers, when $a = 0$ and $b > 0$

④ not defined (no solution), when $a = 0$ and $b \leq 0$

For $ax < b$,

if $a = 0$ then $0 \cdot x < b$

$\Rightarrow \begin{cases} \text{① true, when } b > 0 \\ \quad \therefore \text{ All real numbers are the solution.} \\ \text{② false, when } b \leq 0 \\ \quad \therefore \text{ Solution does not exist (no solution).} \end{cases}$

changed direction

$-ax < b$

$\Rightarrow$ ① if $a > 0$, then $-a < 0$ $\therefore x > -\dfrac{b}{a}$

    ② if $a < 0$, then $-a > 0$ $\therefore x < -\dfrac{b}{a}$

unchanged direction

Find k.

(1) The solution of $ax < b$ is $x < k$.

$\Rightarrow$ Since the direction of the inequality symbol of the solution is unchanged, $a > 0$.

From $ax < b$, $x < \dfrac{b}{a}$ $\therefore k = \dfrac{b}{a}$

(2) The solution of $ax < b$ is $x > k$.

$\Rightarrow$ Since the direction of the inequality symbol of the solution is changed, $a < 0$.

From $ax < b$, $x > \dfrac{b}{a}$ $\therefore k = \dfrac{b}{a}$

5. Steps for Solving Word Problems

Step 1. Remove parentheses, using the distributive property.

Step 2. If there are fractions or decimals for coefficients of variables, change all the coefficients to integers by multiplying a proper number to both sides of the inequality symbol.

Step 3. Isolate the variables and numbers on each side of the inequality symbol.

Step 4. Simplify both sides of the symbol

for any $a (\neq 0)$, $ax < b \, (\, ax \leq b\,)$ or $ax > b \, (\, ax \geq b\,)$

Step 5. Find the solution .

$a > 0 \Rightarrow -a < 0$

$a < 0 \Rightarrow -a > 0$

2-2 Graphing Linear Inequalities with One Variable

1. Solutions on a Number Line

For the linear inequality $ax < b$ or $ax \leq b$ with $a > 0$, the solution is given by the simpler inequality:

$$x < \frac{b}{a} \quad \text{or} \quad x \leq \frac{b}{a}.$$

This is because the solution has an infinite number of values and cannot be expressed in a simpler form. The usual representation of this solution is given by its graph on a number line, using open points and closed points.

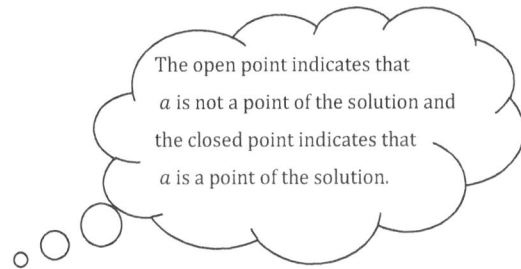

> The open point indicates that a is not a point of the solution and the closed point indicates that a is a point of the solution.

(1) $x < a$

open point

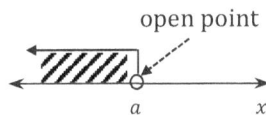

The solution includes an infinite number of values less than a.

(2) $x > a$

open point

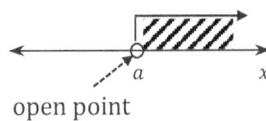

The solution includes an infinite number of values greater than a.

(3) $x \leq a$

closed point

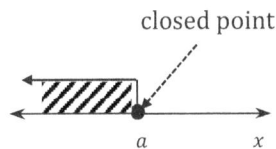

The solution includes an infinite number of values less than or equal to a.

(4) $x \geq a$

closed point

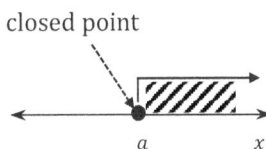

The solution includes an infinite number of values greater than or equal to a.

2. Compound (Combined) Inequalities

A compound inequality consists of two inequalities joined by "and" or "or".

For $m < n$,

(1) $m < x < n \Rightarrow m < x$ and $x < n$

(2) $x > n$ or $x \leq m$

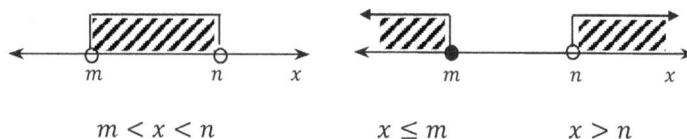

$$m < x < n \qquad\qquad x \leq m \qquad x > n$$

To solve this compound inequality, isolate the variable between the inequality symbols, or isolate the variable in each inequality.

Examples

(1) $1 < 2x - 3 \leq 5 \Rightarrow 1 + 3 < 2x - 3 + 3 \leq 5 + 3$

$\qquad\qquad \Rightarrow 4 < 2x \leq 8$

$\qquad\qquad \Rightarrow \dfrac{4}{2} < x \leq \dfrac{8}{2}$

$\qquad\qquad \Rightarrow 2 < x \leq 4 \Rightarrow x \leq 4$ and $x > 2$

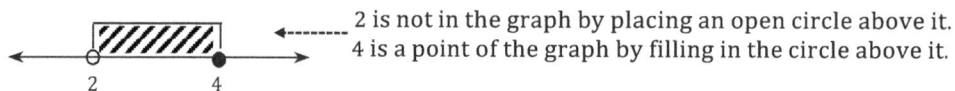

2 is not in the graph by placing an open circle above it.
4 is a point of the graph by filling in the circle above it.

(2) $3x + 4 \leq 2 + 2x$ or $-2x < x + 3$

$\qquad \Rightarrow x \leq -2$ or $3x > -3$

$\qquad \Rightarrow x \leq -2$ or $x > -1$

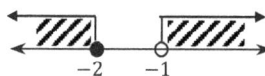

3. Linear Inequalities with Absolute Values

$$|a| = \begin{cases} a, & a \geq 0 \\ -a, & a < 0 \end{cases}$$

For $a > 0$, $|x| > a \Rightarrow x > a$ or $-x > a$

Since $-x > a$ is equivalent to $x < -a$, $|x| > a$ is equivalent to ($x > a$ or $x < -a$).

Similarly, $|x + b| > a$ is equivalent to ($x + b > a$ or $x + b < -a$).

Therefore,

(1) $|x| < a \quad \Leftrightarrow \quad -a < x < a$

(2) $|x + b| < a \quad \Leftrightarrow \quad -a < x + b < a \quad$ or $\quad -a - b < x < a - b$

(3) $|cx + b| < a \quad \Leftrightarrow \quad -a < cx + b < a \quad$ or $\quad -a - b < cx < a - b$

$$\text{or} \quad \frac{-a-b}{c} < x < \frac{a-b}{c}, \; c > 0$$

Examples

(1) $2 < |x - 1| < 3$

① When $x - 1 \geq 0$ ($x \geq 1$),

$2 < |x - 1| < 3 \Rightarrow 2 < x - 1 < 3 \Rightarrow 2 + 1 < x < 3 + 1 \Rightarrow 3 < x < 4$

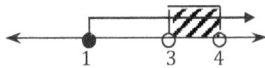

So, $3 < x < 4$

② When $x - 1 < 0$ ($x < 1$),

$2 < |x - 1| < 3 \Rightarrow 2 < -(x - 1) < 3 \Rightarrow 2 - 1 < -x < 3 - 1 \Rightarrow 1 < -x < 2$

$$\Rightarrow -2 < x < -1$$

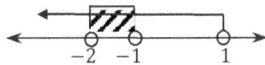

So, $-2 < x < -1$

Therefore, $3 < x < 4$ or $-2 < x < -1$

(2) $|x + 2| + |x - 3| < 10$

Since $x + 2 = 0 \Rightarrow x = -2$ and $x - 3 = 0 \Rightarrow x = 3$,

consider the three cases, $x < -2$, $-2 \leq x < 3$, and $x \geq 3$.

① Case 1 When $x < -2$,

$|x + 2| + |x - 3| < 10 \Rightarrow -(x + 2) - (x - 3) < 10$

$\Rightarrow -2x < 9$

$\Rightarrow x > -\frac{9}{2}$

Since $x < -2$, $-\frac{9}{2} < x < -2$

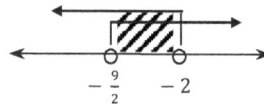

② Case 2 When $-2 \leq x < 3$,

$|x + 2| + |x - 3| < 10 \Rightarrow (x + 2) - (x - 3) < 10$

$\Rightarrow 0 \cdot x < 5$; always true

$\therefore -2 \leq x < 3$

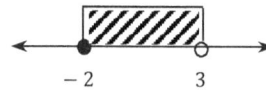

③ Case 3 When $x \geq 3$,

$|x + 2| + |x - 3| < 10 \Rightarrow (x + 2) + (x - 3) < 10$

$\Rightarrow 2x < 11$

$\Rightarrow x < \frac{11}{2}$

Since $x \geq 3$, $3 \leq x < \frac{11}{2}$

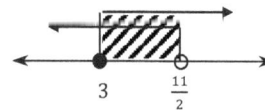

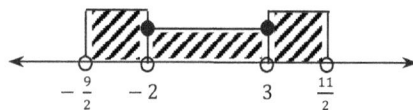

Therefore, the sum of all three intervals is $-\frac{9}{2} < x < \frac{11}{2}$.

Exercises

#1. Express each statement as an inequality.

(1) a is less than -3

(2) a is greater than or equal to 2

(3) a is greater than -1 and less than or equal to 1

(4) 3 more than twice a is greater than half of a

(5) 4 less than three time a is greater than or equal to a plus 2

(6) a is not greater than 0

#2. Solve the following inequalities

(1) $x - 5 > 6$

(2) $x + 4 > 0$

(3) $6x > 3$

(4) $2x + 3 > 7$

(5) $3x - 4 > x + 3$

(6) $x + 5 > 3x$

(7) $-2x - 5 \leq 7$

(8) $-\frac{1}{3}x - 1 \leq 8$

(9) $-2x > 4$

(10) $-3x + 4 < -2x$

(11) $2x > 2(x + 3)$

(12) $5x - (7x - 6) \geq 3$

(13) $3x - (8x + 5) \leq 2$

(14) $2.5x - 1.5 > 3.5x + 4.5$

(15) $2(x + 1) - \frac{8x+1}{3} < 4$

(16) $\frac{4}{3}x - 4\left(\frac{1}{3}x + 2\right) > -1$

(17) $\frac{5x-3}{4} \geq x - \frac{5x+1}{3}$

(18) $0.3 - 0.2x < 0.4x - 0.1$

(19) $3 - 2ax < -3$ for $a < 0$

(20) $-ax - 1 \leq 2$ for $a < 0$

(21) $2ax > -a$ for $a < 0$

(22) $3x < 3x + 4$

#3. Solve the following inequalities. Then draw the solution on a number line

(1) $2x - 4 > 4$

(2) $-3x \leq \frac{x-1}{2} - 3$

(3) $-\frac{x}{4} \geq 2$

(4) $0.3x - \frac{1+x}{2} < -\frac{2}{5}$

#4. Express the range of x as an inequality.

(1) Three times x minus 5 is greater than five times x plus 2

(2) Two times the difference of x and 3 is less than or equal to three times the sum of $2x$ and 2

#5. Express the range of x for the following expression when $-1 \leq x \leq 1$.

(1) $2x + 1$

(2) $-3x - 2$

(3) $\frac{1}{4}x - 3$

#6. Let $y = \frac{4-2x}{3}$.

(1) Find the range of y when $1 < x < 5$.

(2) Find the range of y when $-3 < x < -1$.

(3) Find the range of x when $2 \leq y \leq 4$.

#7. Find the sum of all positive integers which satisfy the inequality

$2(1 - x) + 6 \geq 3(x - 3) - 5$.

#8. The sum of three consecutive integers is greater than or equal to 69. Find the three integers with the smallest sum.

#9. How many positive integers satisfy the following inequalities?

(1) $3\left(\frac{1}{2}x - 1\right) < x + 2$

(2) $0.3(2 - x) \geq 0.1x - 0.2$

(3) $\frac{x+3}{2} - \frac{2-x}{3} < 1$

#10. Only 3 positive integers satisfy the inequality $3x - k \leq \frac{5-x}{2}$. Find the range of k.

#11. Find the constant k if

(1) The inequality $\frac{1}{2}x - \frac{k}{3} < -1$ has the solution $x < 2$.

(2) The inequality $\frac{kx}{4} - \frac{1}{2} > 1$ has the solution $x < -1$.

(3) The inequality $\frac{2-kx}{5} - 2 \leq \frac{x}{2} + 1$ has the solution $x \leq -4$.

(4) Two inequalities $2(1 - 2x) - 3 \leq x - 5$ and $\frac{3k-2x}{3} \leq x + 2k$ have the same solution.

(5) The inequality $2 - kx < 2x + k$ has no solution.

(6) The inequality $-2kx + 5 > 6$ has the solution $x > 2$.

(7) $1 - 5x \leq 2x - 5k$ has -2 as a minimum value of the solution.

(8) The inequality $x - (3 + \frac{k}{2} > 2x + k$ has no positive solution.

#12. The inequality $(-a + 2b)x + b - 3a \leq 0$ has the solution $x \leq -1$. Find the solution for the inequality $(a - b)x + a - 2b > 0$, where $b > 0$.

#13. Solve the following inequality for x and graph the solution

(1) $|x - 2| \leq 0$

(2) $|3x + 9| > 0$

(3) $|x + 4| < 0$

(4) $|-2x + 1| + 3 \leq 6$

(5) $0 < |2 - 4x| < 8$

(6) $2 < |x + 1| < 3$

#14. At the store there is a bucketful of apples and peaches. An apple is worth 25¢ and a peach is worth 50¢. You want to buy 10 pieces of fruit. What is the maximum number of peaches you can buy with less than $10?

#15. Nichole wants to make a salt solution that is at most 10% salt by adding water to 50 ounces of a 15% salt solution. What is the amount of water she can add?

#16. Richard goes hiking in the mountain. He goes up the trail at a speed of 3 miles per hour and down the same trail at a speed of 4 miles per hour, while hiking for no longer than 2 hours. Find the maximum distance he can hike.

#17. Nichole plans to take a 3 miles walk in less than $\frac{1}{2}$ hour. She walks at a speed of 3 miles per hour at the beginning , then runs at a speed of 9 miles per hour for the rest. How far does she walk?

#18. Richard needs to produce a salt solution that is at least 12% salt after mixing 30 ounces of a 5% salt solution with a 15% salt solution. How many more ounces of a 15% salt solution must be needed?

#19. Nichole's last scores on three math tests were 93, 87, and 89 . When she takes her next test, she wants to have a total average of at least 92 points for all four tests. What score does she need to get on her next test?

#20. Richard is 14 years old and his dad is 48 years old. In how many years will his dad's age become less than twice Richard's age?

#21 . The lengths of three sides of a triangle are x, $x + 2$, and $x + 3$, where x is a positive integer. Find the smallest length of the triangle.

#22. The shaded area is, at most, 66 square inches. Find the smallest integer for x.

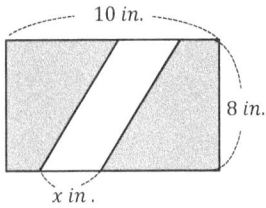

#24. Suppose one boy can complete a task in 4 hours and one girl can complete the same task in 6 hours. A group of 5 boys and girls try to complete the task in 1 hour. Find the minimum number of boys to complete the task.

#23. There is a big sale going on at a book store. Nichole buys a book that is 20% off. But her $20 bill is not enough to buy it. What is the price range of the book?

Algebra

**Part II
Expressions**

Chapter 3

**Monomials and
Polynomials
Level II**

CHAPTER 3

Chapter 3   Monomials and Polynomials

Chapter 3. Monomials and Polynomials

3-1 Exponents

1. Definition

$a^n = n$ repeated factors of a

(1) $a \cdot a \cdot a \cdots a = a^n$: *Exponent form*

(2) *Base* : The number or letter to multiply by itself

(3) *Exponent (Power)* : The number of times to multiply the base by itself

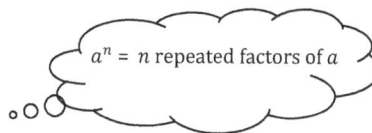

Example

$$3 \cdot 3 \cdot 3 \cdot 3 = 3^4 : \text{3 is the base and 4 is the exponent (or power)}$$

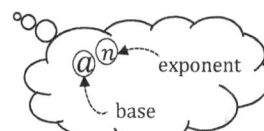

a^{n} exponent
base

2. Rules of Exponents

$1 = 1^1$
Any number to the first power is equal to itself.
$2 = 2^1 \; ; \; 3 = 3^1 \; ; \; \cdots$

For any real number $a(\ne 0)$ and positive integers m and n,

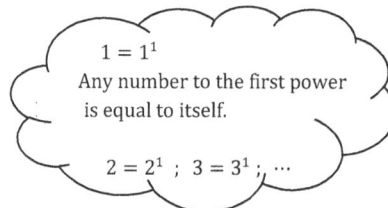

---

(1) Addition of Exponents

$$a^m \cdot a^n = \underbrace{a \cdot a \cdots a \cdot a}_{m \text{ times}} \cdot \underbrace{a \cdot a \cdots a \cdot a}_{n \text{ times}} = a^{m+n}$$

(2) Multiplication of Exponents

$$(a^m)^n = \underbrace{a^m \cdot a^m \cdots a^m \cdot a^m}_{n \text{ times}} = a^{\overbrace{m+m+\cdots+m}^{n \text{ times}}} = a^{mn}$$

(3) Division of Exponents

$$a^m \div a^n = \frac{a^m}{a^n} = \frac{\overbrace{a \cdots a}^{m}}{\underbrace{a \cdots a}_{n}} = \begin{cases} a^{m-n}, & m > n \\ 1, & m = n \\ \frac{1}{a^{n-m}}, & m < n \end{cases}$$

Examples

$$a^3 \div a^2 = \frac{a^3}{a^2} = \frac{a \cdot a \cdot a}{a \cdot a} = a = a^{3-2}$$

$$a^3 \div a^3 = \frac{a^3}{a^3} = \frac{a \cdot a \cdot a}{a \cdot a \cdot a} = 1$$

$$a^2 \div a^3 = \frac{a^2}{a^3} = \frac{a \cdot a}{a \cdot a \cdot a} = \frac{1}{a} = \frac{1}{a^{3-2}}$$

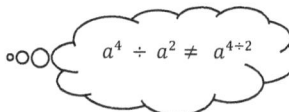

$a^4 \div a^2 \ne a^{4 \div 2}$

Note : When you move the exponential terms across the fraction bar, move the term with the smaller exponent.

For example,

$$a^3 \div a^2 = \frac{a^3}{a^2} = \frac{a^3 \cdot a^{-2}}{1} = a^1 = a$$

$$a^2 \div a^3 = \frac{a^2}{a^3} = \frac{1}{a^3 \cdot a^{-2}} = \frac{1}{a^1} = \frac{1}{a}$$

$$a^{-2} \div a^{-3} = \frac{a^{-2}}{a^{-3}} = \frac{a^{-2} \cdot a^3}{1} = \frac{a^1}{1} = a^1 = a$$

$$a^{-3} \div a^{-2} = \frac{a^{-3}}{a^{-2}} = \frac{1}{a^{-2} \cdot a^3} = \frac{1}{a^1} = \frac{1}{a}$$

3. Distributive Property of Exponents

(1) $(ab)^m = \underbrace{(ab) \cdot (ab) \cdots (ab) \cdot (ab)}_{m \text{ times}}$

$= \underbrace{(a \cdot b) \cdot (a \cdot b) \cdots (a \cdot b) \cdot (a \cdot b)}_{m \text{ times}} = \underbrace{a \cdot a \cdots a \cdot a}_{m \text{ times}} \cdot \underbrace{b \cdot b \cdots b \cdot b}_{m \text{ times}} = a^m \cdot b^m$

(2) $(\frac{a}{b})^m = \underbrace{(\frac{a}{b}) \cdot (\frac{a}{b}) \cdots (\frac{a}{b}) \cdot (\frac{a}{b})}_{m \text{ times}} = \frac{\overbrace{a \cdots a}^{m}}{\underbrace{b \cdots b}_{m}} = \frac{a^m}{b^m}$

Example

$$\left(\frac{a}{b}\right)^2 = \left(\frac{a}{b}\right) \cdot \left(\frac{a}{b}\right) = \frac{a \cdot a}{b \cdot b} = \frac{a^2}{b^2}$$

Note :

- $\left(-3\frac{a}{b}\right)^2 = (-3)^2 \cdot \left(\frac{a}{b}\right)^2 = 9\frac{a^2}{b^2}$
- $(a^l b^m)^n = a^{ln} b^{mn}$, positive integers $l, m, n, a \neq 0, b \neq 0$
- $\left(\frac{a^l}{b^m}\right)^n = \frac{a^{ln}}{b^{mn}}$

4. Expanding Exponents

(1) $a^0 = 1$

(2) $a^{-m} = \dfrac{1}{a^m}$

(3) $a^{m+1} - a^m = a^m(a-1)$

$1 = a^2 \div a^2 = a^{2-2} = a^0$

$1^0 = 1$; $2^0 = 1$; $3^0 = 1$

But $0^0 \neq 1$ ($\because 0^0$ is undefined.)

$a^{m+1} = a^m \cdot a^1 = a^m \cdot a$

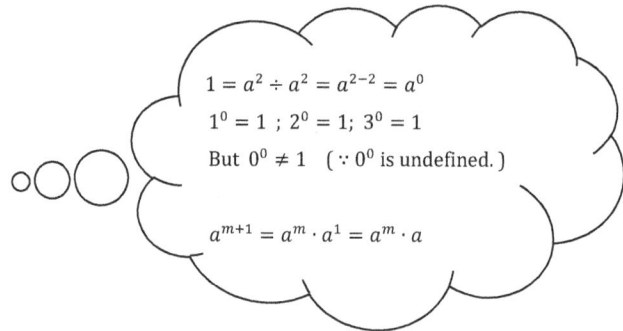

Negative exponents

1) $a^{-m} = \dfrac{1}{a^m}$

For example, $\dfrac{a^2}{a^5} = a^{2-5} = a^{-3}$, by the same base quotient rule, and $\dfrac{a^2}{a^5} = \dfrac{a \cdot a}{a \cdot a \cdot a \cdot a \cdot a} = \dfrac{1}{a \cdot a \cdot a} = \dfrac{1}{a^3}$

$\therefore \ a^{-3} = \dfrac{1}{a^3}$

2) $\dfrac{1}{a^{-m}} = a^m$

$\because$ Since $a^{-m} = \dfrac{1}{a^m}$,

$\dfrac{1}{a^{-m}} = \dfrac{1}{\frac{1}{a^m}}$, by substituting $\dfrac{1}{a^m}$ for a^{-m}

$= \dfrac{\frac{1}{1}}{\frac{1}{a^m}} = \dfrac{1 \cdot u^m}{1 \cdot 1} = \dfrac{u^m}{1} = a^m$

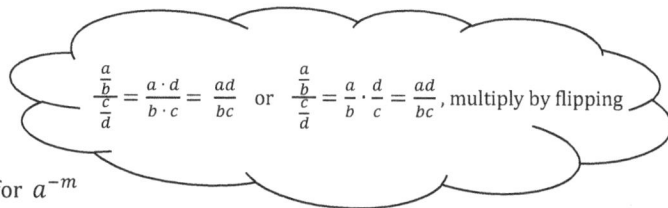

$\dfrac{\frac{a}{b}}{\frac{c}{d}} = \dfrac{a \cdot d}{b \cdot c} = \dfrac{ad}{bc}$ or $\dfrac{\frac{a}{b}}{\frac{c}{d}} = \dfrac{a}{b} \cdot \dfrac{d}{c} = \dfrac{ad}{bc}$, multiply by flipping

3) $\dfrac{1}{a^x} = a^{-x}$; $\dfrac{1}{a^{-x}} = a^x$; $\dfrac{1}{a^{-m}} = a^m$ (opposite power)

1 over any term raised to a power (exponent) is the same as that term raised to the opposite power (exponent) .

4) $a^m \cdot a^{-m} = a^{m-m} = a^0 = 1$ or $a^m \cdot a^{-m} = a^m \cdot \dfrac{1}{a^m} = 1$

a^{-m} is the multiplicative inverse (or reciprocal) of a^m .

3-2 Multiplying and Dividing Monomials

A monomial is an expression consisting of only one term which is product of numbers and variables.

Solving steps :

> $3x^2$: monomial of degree 2 in x
>
> $2x$: monomial of degree 1 in x

Step 1. Solve parentheses using the rules of exponents

Step 2. $\div a \ \rightarrow \ \times \frac{1}{a}$ (reciprocal)

Step 3. Determine the sign : number of $(-)$ sign is even $\Rightarrow (+)$

  number of $(-)$ sign is odd $\Rightarrow (-)$

Step 4. Calculate the exponents

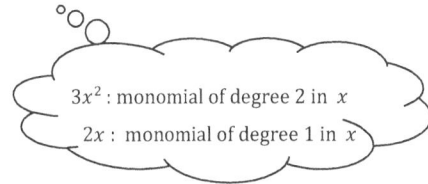

Examples

1. $-2a \cdot 3ab = (-2 \cdot 3) \cdot (a \cdot ab) = -6 \cdot a^2 b = -6\,a^2 b$

2. $-2a \div 3ab = -2a \cdot \dfrac{1}{3ab} = \dfrac{-2a}{3ab} = \dfrac{-2}{3} \cdot \dfrac{a}{ab} = -\dfrac{2}{3} \cdot \dfrac{1}{b} = -\dfrac{2}{3b}$

3. $3a \cdot a^2 \div (3a^2)^3 = 3a \cdot a^2 \div 3^3 a^6 = 3a \cdot a^2 \cdot \dfrac{1}{3^3 a^6} = \dfrac{3}{3^3} \cdot \dfrac{a^3}{a^6} = \dfrac{1}{3^2} \cdot \dfrac{1}{a^3} = \dfrac{1}{9a^3}$

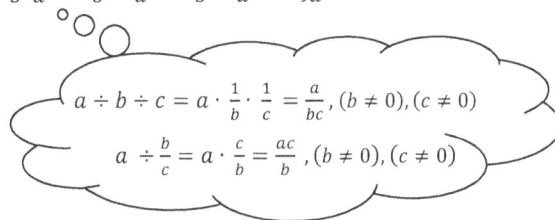

> $a \div b \div c = a \cdot \dfrac{1}{b} \cdot \dfrac{1}{c} = \dfrac{a}{bc}, (b \neq 0), (c \neq 0)$
>
> $a \div \dfrac{b}{c} = a \cdot \dfrac{c}{b} = \dfrac{ac}{b}, (b \neq 0), (c \neq 0)$

3-3   Polynomials

A polynomial is an expression of two or more terms combined by addition and/or subtraction.

1. Adding and Subtracting Polynomials

Steps to Solve:

Step 1. Remove the parentheses ( ) or other enclosure marks (brace{ }, bracket[ ], absolute value sign | |, etc.) in order from the innermost enclosure marks to the outermost ones.

Step 2. Regroup the like terms.

Step 3. Simplify by combining the expressions.

Note : (1) *A polynomial is an expression which is the sum of monomials.*

For example, $3x^2 + 2x,\ 2x - 1$

(2) *The terms in a polynomial are ordered from the greatest exponent to the least (decreasing degree in variable).*

$a_n x^n + a_{n-1} x^{n-1} + \cdots + a_2 x^2 + a_1 x^1 + a_0$

For example, $x^2 + 2x^3 + x + 4 \rightarrow 2x^3 + x^2 + \boxed{x} + \boxed{4}$

$\cdots\cdots 4x^0$

x^1

$x + (y - z) = x + y - z$
$x - (y - z) = x - y + z$

$\dfrac{a}{b} + \dfrac{c}{d} = \dfrac{ad + bc}{bd}$
(bd is the common denominator)

Examples

1. $(2x + 3y) - (x + 2y) = 2x + 3y - x - 2y = (2x - x) + (3y - 2y) = x + y$

2. $\dfrac{x+2y}{2} - \dfrac{2x-y}{3} = \dfrac{3(x+2y)}{6} - \dfrac{2(2x-y)}{6} = \dfrac{3x+6y-4x+2y}{6} = \dfrac{(3x-4x)+(6y+2y)}{6} = \dfrac{-x+8y}{6}$

3. $2x - [3y - \{x - (2x - 3y) + 2y\} - 5] = 2x - [3y - \{x - 2x + 3y + 2y\} - 5]$

$= 2x - [3y - x + 2x - 3y - 2y - 5]$

$= 2x - 3y + x - 2x + 3y + 2y + 5$

$= (2x + x - 2x) + (-3y + 3y + 2y) + 5$

$= x + 2y + 5$

4. $2(x^2 - 3x + 5) - 3(2x^2 - x + 3) = 2x^2 - 6x + 10 - 6x^2 + 3x - 9$

$= (2x^2 - 6x^2) + (-6x + 3x) + (10 - 9)$

$= -4x^2 - 3x + 1$

Note : *Linear form : when the degree of $ax + b$ is 1.*

Quadratic form : when the degree of $ax^2 + bx + c$ is 2.

Cubic form : when the degree of $ax^3 + bx^2 + cx + d$ is 3.

For addition,
Commutative property : $a + b = b + a$
Associative property : $(a + b) + c = a + (b + c)$

2. Multiplying and Dividing a Polynomial by a Monomial

To multiply or divide a polynomial by a monomial, use the distributive property and the rules of exponents.

$$(1)\ a \cdot (b + c) = a \cdot b + a \cdot c \qquad (a + b) \cdot c = a \cdot c + b \cdot c$$

$$(2)\ (a + b) \div c = (a + b) \cdot \frac{1}{c} = \left(a \cdot \frac{1}{c}\right) + \left(b \cdot \frac{1}{c}\right) = \frac{a}{c} + \frac{b}{c}$$

$a^m \cdot a^n = a^{m+n}$: same base

$\dfrac{a^m}{a^n} = a^{m-n}$: same base

$\dfrac{a+b-c}{d} = \dfrac{a}{d} + \dfrac{b}{d} - \dfrac{c}{d}$: same denominator

For Multiplication,

Commutative property : $a \cdot b = b \cdot a$

Associative property   : $a \cdot (b \cdot c) = (a \cdot b) \cdot c$

Distributive property : $a \cdot (b + c) = a \cdot b + a \cdot c$

$(b + c) \cdot a = b \cdot a + c \cdot a$

Steps to Solve :

Step 1.   Remove the parentheses ( ) or other enclosure marks (brace{ }, bracket[ ], absolute

value sign | |, etc.) in order from the innermost enclosure marks to the outermost marks.

Step 2.   Exponents(Powers)

Step 3.   ×, ÷ (Compute in order from left to right, no matter which operation comes first)

Step 4.   +, − (Compute in order from left to right, no matter which operation comes first)

Note : ① $(a \times b) \div c = a \times (b \div c)$

$\because (a \times b) \div c = \dfrac{a \times b}{c} = \dfrac{ab}{c}$ and $a \times (b \div c) = a \times \dfrac{b}{c} = \dfrac{ab}{c}$

② $(a \div b) \times c \neq a \div (b \times c)$

③ $a \div b \times c = (a \div b) \times c \neq a \div (b \times c)$

for multiplying and dividing, calculate from left to right by order.

3. Multiplying Polynomials

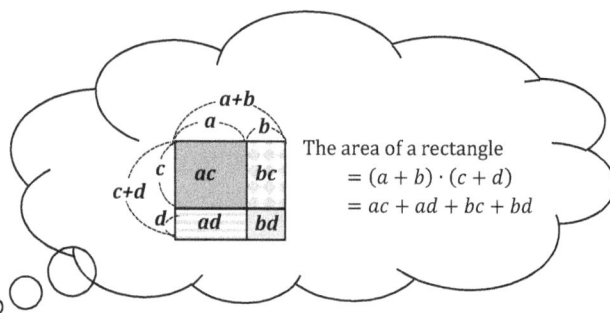

The area of a rectangle
$= (a + b) \cdot (c + d)$
$= ac + ad + bc + bd$

$$(a + b) \cdot (c + d) = ac + ad + bc + bd$$

Note

- $(a + b) \cdot (c + d) = a(c + d) + b(c + d) = ac + ad + bc + bd$

- $(x + a)(x + b) = x^2 + (a + b)x + ab$

- $(ax + b)(cx + d) = acx^2 + (ad + bc)x + bd$

- $(a + b + 1)(a + b - 1) = (A + 1)(A - 1), \quad letting \ a + b = A$

$$= A^2 - A + A - 1 = A^2 - 1 = (a + b)^2 - 1$$

$$= (a + b)(a + b) - 1 = a^2 + ab + ba + b^2 - 1$$

$$= a^2 + 2ab + b^2 - 1$$

- $(a + 1)(a + 2)(a + 3)(a + 4) = (a + 1)(a + 4)(a + 2)(a + 3) \quad ; regroup \ to \ get \ a^2 + 5a$

$$= (a^2 + 5a + 4)(a^2 + 5a + 6)$$

$$= (A + 4)(A + 6), \quad letting \ a^2 + 5a = A$$

$$= A^2 + 10A + 24$$

$$= (a^2 + 5a)^2 + 10(a^2 + 5a) + 24, \quad replace \ A \ as \ a^2 + 5a$$

$$= a^4 + 10a^3 + 35a^2 + 50a + 24$$

If the number of $(-)$ signs in the product is even, then $(+)$.
$(-) \cdot (-) = (+) \ ; \ (-) \cdot (-) \cdot (-) \cdot (-) = (+)$

If the number of $(-)$ signs in the product is odd, then $(-)$.
$(-) \cdot (+) = (-) \ ; \ (-) \cdot (-) \cdot (-) = (-)$

4. Formulas

(1) $\boxed{(a + b)^2 = a^2 + 2ab + b^2}$

$\because (a + b)^2 = (a + b)(a + b) = a^2 + ab + ba + b^2 = a^2 + 2ab + b^2$

(2) $\boxed{(a - b)^2 = a^2 - 2ab + b^2}$

$\because (a - b)^2 = (a - b)(a - b) = a^2 - ab - ba + b^2 = a^2 - 2ab + b^2$

(3) $\boxed{a^2 + b^2 = (a + b)^2 - 2ab = (a - b)^2 + 2ab}$

$(a + b)^2 \neq a^2 + b^2$
$(a - b)^2 \neq a^2 - b^2$

(4) $\boxed{a^2 - b^2 = (a + b)(a - b)}$

Note :

$$(F + B)^2 = F^2 + \mathbf{2}FB + B^2$$
$$(F - B)^2 = F^2 - \mathbf{2}FB + B^2$$
$$F^2 - B^2 = (F + B)(F - B)$$

$$(a + b)^2 = \boxed{a^2} + \underline{2ab} + \boxed{b^2}$$
$$front^2 \qquad back^2$$
$$2 \times front \times back$$

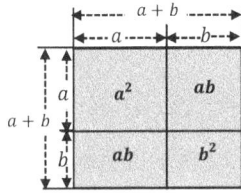

$$(a + b)^2 = (a + b)(a + b) = a^2 + 2ab + b^2$$

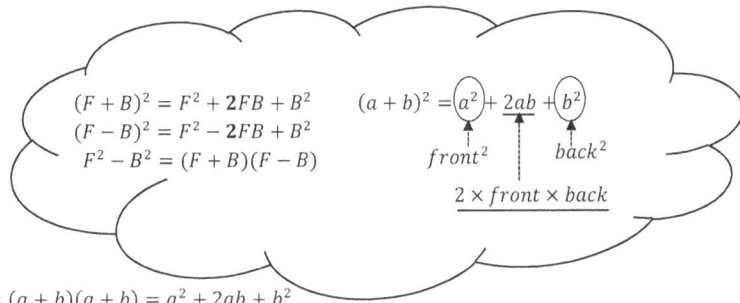

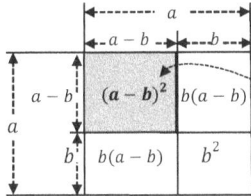

$$(a - b)^2 = a^2 - 2b(a - b) - b^2 = a^2 - 2ba + 2b^2 - b^2 = a^2 - 2ab + b^2$$

Examples

$$101^2 = (100 + 1)^2 = 100^2 + 2 \cdot 100 \cdot 1 + 1^2 = 10201$$
$$99^2 = (100 - 1)^2 = 100^2 - 2 \cdot 100 \cdot 1 + 1^2 = 9801$$
$$101 \cdot 99 = (100 + 1)(100 - 1) = 100^2 - 1^2 = 9999$$

Expanding Formulas

(1) $\left(a + \dfrac{1}{a}\right)^2 = a^2 + 2 \cdot a \cdot \dfrac{1}{a} + \dfrac{1}{a^2} = a^2 + \dfrac{1}{a^2} + 2$

(2) $\left(a - \dfrac{1}{a}\right)^2 = a^2 - 2 \cdot a \cdot \dfrac{1}{a} + \dfrac{1}{a^2} = a^2 + \dfrac{1}{a^2} - 2$

(3) $a^2 + \dfrac{1}{a^2} = \left(a + \dfrac{1}{a}\right)^2 - 2 \cdot a \cdot \dfrac{1}{a} = \left(a + \dfrac{1}{a}\right)^2 - 2$

$\quad\ a^2 + \dfrac{1}{a^2} = \left(a - \dfrac{1}{a}\right)^2 + 2 \cdot a \cdot \dfrac{1}{a} = \left(a - \dfrac{1}{a}\right)^2 + 2$

(4) $a^2 - \dfrac{1}{a^2} = \left(a + \dfrac{1}{a}\right)\left(a - \dfrac{1}{a}\right)$

(5) $\left(a + \dfrac{1}{a}\right)^2 = \left(a - \dfrac{1}{a}\right)^2 + 4$

(6) $\left(a - \dfrac{1}{a}\right)^2 = \left(a + \dfrac{1}{a}\right)^2 - 4$

(7) $(a + b)^2 = (a - b)^2 + 4ab$

(8) $(a - b)^2 = (a + b)^2 - 4ab$

Note $\quad (a + b + c)^2 = ((a + b) + c)^2 = (A + c)^2, \ \ letting \ \ a + b = A$

$\qquad\qquad\qquad = A^2 + 2Ac + c^2$

$\qquad\qquad\qquad = (a + b)^2 + 2(a + b)c + c^2, \ \ replace \ A \ as \ a + b$

$\qquad\qquad\qquad = a^2 + 2ab + b^2 + 2ac + 2bc + c^2$

$\qquad\qquad\qquad = a^2 + b^2 + c^2 + 2ab + 2bc + 2ac \,, \ rearranging \ the \ terms$

5. Special Equalities

(1) The value of an algebraic expression $2x + 3y$ when $x = 3$ and $y = 4$ is
$(2 \cdot 3) + (3 \cdot 4) = 6 + 12 = 18$.

(2) When $x = y + 1$, $x + y = (y + 1) + y = 2y + 1$, replacing x with $y + 1$

(3) Solve an equation which contains two or more variables for a given variable by isolating the variable to one side of the equation.

Example

Solve the equation $ax + by = c$ for the variables x and y separately.

For x, $ax = -by + c$; $x = -\dfrac{b}{a}y + \dfrac{c}{a}$

For y, $by = -ax + c$; $y = -\dfrac{a}{b}x + \dfrac{c}{b}$

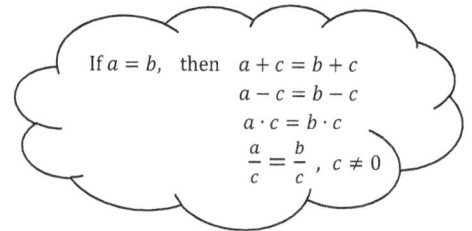

If $a = b$, then $a + c = b + c$
$a - c = b - c$
$a \cdot c = b \cdot c$
$\dfrac{a}{c} = \dfrac{b}{c}$, $c \neq 0$

Exercises

1. Simplify each expression.

(1) $a^2 \cdot a^3 \cdot a^4$

(2) $x^3 \cdot y^2 \cdot x^4 \cdot y \cdot z$

(3) $(2^3 xy^2 z^3)^2$

(4) $(x^3)^2 \cdot (x^4)^3$

(5) $((-x)^2)^3 \cdot ((-x)^3)^2$

(6) $(-a^2 b^3)^5$

(7) $-3xy^2 (-2x^2 yz^3)^3$

(8) $\left(-\dfrac{x}{y^2}\right)^2$

(9) $\dfrac{a^2 a^3}{(-a)^4}$

(10) $\left(\dfrac{2}{3}a^2\right)^2 \cdot \left(\dfrac{3}{4}a^3\right)^2$

(11) $(-a^2 b)^3 \div (-a)^3 \cdot (ab^2)^2$

(12) $\left(\dfrac{2}{3}\right)^{-3}$

(13) $\left(\dfrac{ab}{a^2 b^3}\right)^2$

(14) $\dfrac{x^3 x^{-4}}{x^2}$

(15) $\dfrac{a^3 b^{-2}}{a^{-4} b^3}$

(16) $\dfrac{2^3 + 2^3 + 2^3}{5^2 + 5^2 + 5^2}$

(17) $3^{2a-1} + 3^{2a-1} + 3^{2a-1}$

(18) $\dfrac{3^4 + 3^5 + 3^6 + 3^7}{3 + 3^2 + 3^3 + 3^4}$

(19) $\dfrac{4^3 + 4^3 + 4^3 + 4^3}{4^3 \cdot 4^3 \cdot 4^3 \cdot 4^3}$

(20) $-3xy^2 \cdot (-2x^2 y)^3 \div (2xy)^2$

(21) $\left(-\dfrac{2}{3}xy^3\right)^3 \div 4x^2 y \cdot \left(-\dfrac{4}{3}x^3 y\right)^2$

(22) $3^{-1} \cdot \left(\dfrac{1}{2}\right)^3 \cdot 3^3$

(23) $8^{a-1} \cdot 2^{3a+1} \div 4^{3a-1}$

2. Find all expressions that are true.

(1) $(a^2)^3 = a^5$

(2) $(-a)^3 \cdot -a^2 = -a^5$

(3) $a^3 \div a^3 = a^1$

(4) $a^4 \div a^3 \cdot a^5 = a^6$

(5) $a^2 + a^3 = a^5$

(6) $a^{-2} \cdot b^{-2} = (ab)^{-2}$

(7) $\left(\dfrac{a}{b^2}\right)^3 = \dfrac{a^3}{b^5}$

(8) $(a^2b)^3 \div -2ab = -\dfrac{1}{2}a^5b^2$

(9) $(a^2)^3 = a^{2^3}$

(10) $\left(\dfrac{3}{x}\right)^2 = \dfrac{1}{x^2}$

(11) $(2a^2)^3 = 6a^6$

3. Find a and b for the following.

(1) $32^3 = (2^a)^3 = 2^b$

(2) $2^{a+3} = 8^3$

(3) $(2^3)^2 \cdot (2^4)^a = 2^{18}$

(4) $(3^b)^3 \div 3^5 = 3^{10}$

(5) $(4^3)^a = 2^{42}$

(6) $24^4 = 2^a \cdot 3^b$

(7) $16^a = 2^{a+3}$

(8) $(2^3)^4 \div 8^3 \cdot (3^3)^2 = 2^a \cdot 3^b$

(9) $(2)^{3a+1} \div (2)^{2a-3} = 4$

(10) $5^a + 5^{a+2} = 3250$

(11) $2^a + 2^{a+2} = 160$

(12) $2^{a-2} = 0.5^{2a-1}$

(13) $(-2x^a)^b = -32x^{15}$

(14) $2^{a+2} = 2^{a+1} + 8$

(15) $(x^3y)^2 \cdot (xy^2)^a \div x^2y^3 = x^by^{13}$

4. $a = 2^{x+1}, b = 3^{x-1}$. Express 6^x using a and b.

5. $10^x = 2$, $10^y = 3$. Simplify $6^{\frac{x-y}{x+y}}$.

6. For a positive integer n, compute $(-1)^{2n+1} \cdot (-1)^{3n-1} \cdot (-1)^{2n-1} \div (-1)^{3n}$

7. Order the following numbers from least to greatest $\ 2^{32}$, 4^{10}, 8^7, $\left(\frac{1}{2}\right)^{-30}$

8. Find the sum of all possible values of a natural number a which satisfies $a^{2a-1} = a^{3a-4}$.

9. For any positive number n, $\ 2^{n+3}(3^{n+1} + 3^{n+2}) = a6^n$. Find the value of a.

10. For a solid with a length of a^3b^4, width of $3ab^2$, and volume of $15\,a^7b^8$, find the height.

11. Find the number of digits in the final value of the following expressions
   (1) $2^4 \cdot 3^2 \cdot 5^5$
   (2) $5^2 \cdot 3^3 \cdot 20 \cdot 6$
   (3) $4^8 \cdot 5^{10}$

12. Simplify each polynomial.

(1) $(2a + 3b) - (a - b)$

(2) $(3a^2 - a + 3) - (-5a + 3)$

(3) $(2a + 3) - (a^2 - 2a + 5)$

(4) $4x - \{-2y + 3x - (2x - y) + 3\} - (x - 3y)$

(5) $\frac{1}{3}x - \frac{2}{3}y - (2x + 3y) - \frac{1}{2}x$

(6) $\frac{x+3y-1}{2} - \frac{2x-y+2}{3}$

(7) $2a - [a^2 - \{3b - (2a - b) + a^2\} - 5]$

13. When an integer a is divided by 5, the remainder is 1. When an integer b is divided by 5, the remainder is 2 . Find the remainder when $a + b$ is divided by 5.

14. a is the coefficient of x^2 and b is the constant of the following polynomials. Find $a - b$.

    (1) $-(2x^2 - 4x + 5) + (3x^2 - x + 1)$

    (2) $\left(\frac{1}{3}x^2 - \frac{1}{2}x + 2\right) - (\frac{1}{2}x^2 - \frac{2}{3}x + 5)$

    (3) $2x^2 - \{3x - (3x^2 + 2x)\} - 2x + 5$

15. You wanted to add the polynomial $-2a^2 + 3a - 4$ to a polynomial A, but you accidentally subtracted the polynomial from A and got $-3a^2 + 5$. Compute the right answer.

16. If you subtract the polynomial $2a^2 - a + 3$ from two times a polynomial A, then you get $-2a^2 - a - 2$. If you add two times the polynomial $2a^2 - a + 3$ to a polynomial A, then you get $4a^2 + a + 2$. Find the value of a satisfying the two conditions.

17. Find the perimeter of the following shapes.

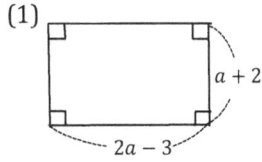

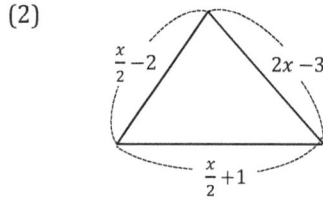

(1)

$a + 2$

$2a - 3$

(2)

$\frac{x}{2} - 2$

$2x - 3$

$\frac{x}{2} + 1$

18. Simplify each polynomial.

(1) $-2x(3x + 4y - 2)$

(2) $x(x^2 - xy + y^2) - y(-x^2 + xy + y^2)$

(3) $(3a + 2b - 4ab) \cdot -\frac{1}{2}a$

(4) $(a^2b - ab^2) \div (-ab)$

(5) $(3a^2b - 2ab^2) \div \left(-\frac{2}{3}ab\right)$

(6) $2a(a-1) - (a^2-1) - 2a(-a+1)$

(7) $-\frac{5}{6}x^2y \cdot \left(\frac{3}{5}xy^2 - 3xy\right)$

(8) $\left(\frac{2}{3}xy^2 - 2x^2y^2\right) \cdot \left(-\frac{3}{2}xy\right) + \left(\frac{4}{3}x^2y - xy^2\right) \div \left(-\frac{3}{2}xy\right)$

(9) $\frac{4x^2 - 2xy}{2xy} - \frac{6xy^2 - 9y^2}{3xy}$

(10) $(6x^2y + 3x^2) \div 3x - (3xy - 9y^2) \div 3y$

(11) $\left(\frac{1}{8}ab - \frac{1}{2}a\right) \cdot 4b - \left(\frac{3}{4}a^2b^2 + a^2b\right) \div 3a$

(12) $\left\{\frac{1}{2}x^2 - \frac{2}{3}(x-3)\right\} + 3\left\{\frac{1}{2}(x-2) - \frac{1}{3}(x^2+3) + 2\right\}$

19. $(2x^2y^3)^a \div 4xy \cdot \frac{1}{2}x^2y = bx^3y^3$. Find the value of $a + b$, where a and b are constants.

20. Find the value of $a + b + c$ for the following, where $a, b,$ and c are constants:

(1) $\left(\frac{4}{3}x^2y - 3xy^2 + 2xy\right) \div \frac{1}{2}xy = ax + by + c$

(2) $\frac{1}{2}(x^2 - 3x + 1) - 2x(x - 1) + 3(4x^2 - 3x - 2) = ax^2 + bx + c$

21. Find the polynomial for each expression.

(1) If a polynomial is multiplied by $2ab$, the result is $\frac{1}{2}a^2b + ab^2 - \frac{1}{3}ab$.

(2) If a polynomial is divided by $3a - 2b$, the quotient is $\frac{1}{4}ab$ and there is no remainder.

22. Find the area of the shaded part in the rectangle. The rectangle has a length of $4a$ and width of $2b$.

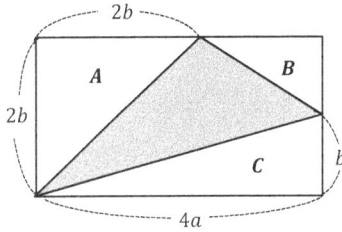

#23. Expand and simplify each polynomial.

(1) $(2x - 5)(x + 3)$

(2) $(2x - 1)(3x^2 - x - 2)$

(3) $\left(x + \frac{1}{3}\right)\left(x - \frac{1}{2}\right)$

(4) $(3 - 2a)(3 + 2a)$

(5) $(3a - 2b)(3a + 2b)$

(6) $(2x + 3)(2x + 3)$

(7) $(-2x + 3)(-2x - 3)$

(8) $(a^3 + b^3)(a^3 - b^3)$

(9) $\left(-4x - \frac{1}{2}\right)^2$

(10) $(x + y - 2)^2$

(11) 102×98

(12) 92×93

(16) $111 \times 109 - 107 \times 113$

(13) 99^2

(17) $(2a + b)^2 - (2a - b)^2$

(14) $(x + 2y + 3z)(x + 2y - 3z)$

(18) $(a - 3)(a + 2)(a - 1)(a + 4)$

(15) $(2x + y - 3)(2x - y + 3)$

24. Find the area of the shaded part of each shape.

(1)

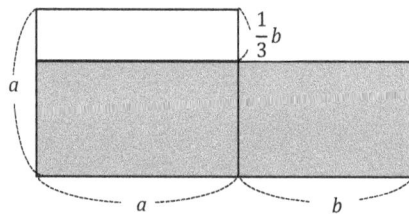

(2)

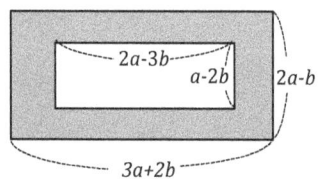

(3)

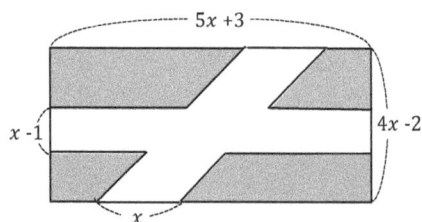

25. Evaluate the polynomial for the variable in each expression.

(1) $3x - 2$ for $x = -2$

(2) $\frac{2}{3}x + 3$ for $x = -1$

(3) $-2x^2 - 3x + 1$ for $x = -3$

(4) $(2x - 2)(-3x + 1)$ for $x = 2$

26. Find the values of the following polynomials

(1) $a^2 + \frac{1}{b^2}$ when $a - \frac{1}{b} = 3$, $\frac{b}{a} = -\frac{1}{3}$

(2) $a^2 + \frac{1}{a^2}$ when $a + \frac{1}{a} = 3$

(3) ab when $a - b = 4$, $a^2 + b^2 = 8$

(4) $\left(a - \frac{1}{a}\right)^2$ when $a + \frac{1}{a} = -3$

(5) $\frac{3a+3b}{2a-4b}$ when $\frac{a-b}{a+b} = \frac{2}{3}$

(6) $\dfrac{b-c}{a} + \dfrac{c-a}{b} - \dfrac{a+b}{c}$ when $a + b - c = 0$, $(abc \neq 0)$

(7) $a^4 + b^4$ when $a - b = 1$, $ab = 2$

(8) $\dfrac{(3a-2b)^2}{(2a+3b)^2}$ when $a : b = 3 : 2$

(9) $\left(\dfrac{2}{3}a^2 - \dfrac{3}{2}b^2\right)\left(-\dfrac{2}{3}a^2 - \dfrac{3}{2}b^2\right)$ when $a = \dfrac{1}{2}$, $b = \dfrac{1}{3}$

(10) $\dfrac{-3a-6ab+3b}{a-3ab-b}$ when $\dfrac{1}{a} - \dfrac{1}{b} = 3$, $ab \neq 0$

(11) $x^2 + \dfrac{9}{x^2} - 3$ when $x^2 - 4x - 3 = 0$

(12) $x^2 + \dfrac{1}{x^2} - 2x + \dfrac{2}{x}$ when $x^2 + 3x - 1 = 0$

(13) $\dfrac{y}{x} + \dfrac{x}{y}$ when $x - y = 3$, $(x + 2)(y - 2) = -6$

(14) $\dfrac{3x+4xy-3y}{x-y}$ when $\dfrac{1}{x} - \dfrac{1}{y} = 2$

(15) $\dfrac{x-2y}{2x+y}$ when $\dfrac{3x+y}{2} = \dfrac{2x-y}{3}$

(16) $\dfrac{a^2-b^2}{(a+b)^2}$ when $(3x + a)(bx - 1) = (3x - 1)^2$

(17) $\dfrac{1}{xyz}$ when $x + \dfrac{1}{y} = 1$, $y + \dfrac{1}{z} = 1$

(18) $(x + 1)(x + 2)(x - 3)(x - 4)$ when $x^2 - 2x - 5 = 0$

27. Evaluate each equation for the specified variable.

(1) $2x - 3y + 6 = 0$ for x

(2) $x = -2y + 3$ for y

(3) $2a = \dfrac{1}{3}(2b - 1)$ for b

(4) $c = \frac{5}{9}(F - 32)$ for F

(5) $a + b : a - b = 3 : 5$ for a

(6) $\frac{1}{a} + \frac{1}{b} = \frac{1}{c}, \ (a \neq 0, b \neq 0, c \neq 0)$ for a

(7) $(2a - b)(a + b) = (a + 3b)(2a - b)$ for a

(8) $a = -\frac{c}{b} + 1, \ b \neq 0, c \neq 0$ for b

28. Find the value of $a + b$, where a and b are constants.

(1) $(5 - 1)(5 + 1)(5^2 + 1)(5^4 + 1) = 5^a - b$

(2) $8(3^2 + 1)(3^4 + 1)(3^8 + 1) = 3^a + b$

(3) $(2x + ay)(x - 3y) = 2x^2 - bxy + 9y^2$

(4) $(x + y)(x - y) - (3y - 2x)(2x - 3y) = \frac{1}{a}x^2 - 12xy - \frac{1}{b}y^2$

29. Find the sum of the coefficients as well as the constants for each polynomial.

(1) $(x + 2y - 3)(x + 2y - 2)$

(2) $(2x - 3y - 5)(2x + 3y - 5)$

(3) $(3x + 2y)(x - 2y) - (x + 3y)(x - 2y)$

Algebra

Part II Expressions

Chapter 4

Systems of Equations Level II

CHAPTER 4

Chapter 4   Systems of Equations

Chapter 4. Systems of Equations

4-1 Systems of Equations

1. Definition

A *linear equation* is an equation in the form of

$$ax + by = c$$, where a, b, and c are constants and $a \neq 0, b \neq 0$.

For example, $2x + 3y - 5 = 0$ or $2x + 3y = 5$.

A *system of equations* is a pair of two or more linear equations with the same variables.

For example, $\begin{cases} x + y = 2 \\ 2x + y = 3 \end{cases}$

For the values of x and y which are satisfying the two linear equations,

the ordered pair (x, y) is called the *solution* for the system of equations.

Note : A = Set of solutions for equation ①
B = Set of solutions for equation ②.

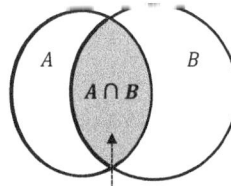

Solutions for the system of equations ① and ②.

Example

For any positive integers x and y, find the solution for the system $\begin{cases} x + y = 3 \cdots\cdots ① \\ 2x + y = 5 \cdots\cdots ② \end{cases}$

For equation ①, the solutions are $(1, 2), (2, 1)$ and for an equation ②, the solutions are $(1, 3), (2, 1)$.

Therefore, the solution for the system of equations ① and ② is $(2, 1)$.

2. Solving Systems of Equations

(1) The Elimination Method

> The *elimination method* is a method for finding a solution by eliminating (removing) one variable by using addition or subtraction.

Note 1 : *If the sign of the coefficient of one variable which is supposed to be removed is different from the sign of the coefficient of the other variable, use addition for the two equations to remove the variable.*
If the sign of the coefficient of one variable which is supposed to be removed is the same as the sign of the coefficient of the other variable, use subtraction for the two equations to remove the variable.

Note 2 : *One or two of the given equations needs to be multiplied by a number in order to make the two equations with the same absolute value for the coefficient of the variable which is supposed to be removed.*

Example 1

$$\begin{cases} 2x + y = 3 & \cdots\cdots \text{①} \\ x - y = 6 & \cdots\cdots \text{②} \end{cases}$$

Step 1.   Add equations ① and ② to remove the variable y. Then solve for x.

$$\begin{array}{r} 2x + y = 3 \quad\cdots\cdots\text{①} \\ +)\ \underline{x - y = 6 \quad\cdots\cdots\text{②}} \\ 3x \quad\ \ = 9 \ ; \ x = 3 \end{array}$$

Step 2.   Substitute the value of x into any one of the given two equations and then solve for y.

$$x - y = 6 \ \Rightarrow \ 3 - y = 6 \ \Rightarrow \ y = -3$$

Step 3.   Find the solution.

$$(x, y) = (3, -3)$$

Example 2

$$\begin{cases} 2x + y = 4 \cdots\cdots \text{①} \\ 3x + y = 2 \cdots\cdots \text{②} \end{cases}$$

Step 1. Subtract equation ② from equation① to remove the variable y.

Then solve for x.

$$2x + y = 4 \cdots\cdots \text{①}$$
$$-)\ \underline{3x + y = 2 \cdots\cdots \text{②}}$$
$$-x\quad\ = 2 \quad ;\ x = -2$$

Step 2. Substitute the value of x into any one of the given two equations.

Then solve for y.

$$2x + y = 4 \ \Rightarrow\ -4 + y = 4 \ \Rightarrow\ y = 8$$

Step 3. Find the solution.

$$(x, y) = (-2,\ 8)$$

Example 3

$$\begin{cases} 2x + 3y = 3 \cdots\cdots \text{①} \\ 3x + 4y = 1 \cdots\cdots \text{②} \end{cases}$$

Step 1. Consider *Note*1 and *Note* 2.

$$6x + 9y = 9 \cdots\cdots \text{①} \times 3$$
$$-)\ \underline{6x + 8y = 2 \cdots\cdots \text{②} \times 2}$$
$$y = 7$$

Step 2. Substitute the value of y into any one of the given two equations.

Then solve for x.

$$2x + 3y = 3 \ \Rightarrow\ 2x + 21 = 3 \ \Rightarrow\ 2x = -18 \ \Rightarrow\ x = -9$$

Step 3. Find the solution.

$$(x, y) = (-9,\ 7)$$

(2) The Substitution Method

> The substitution method is a method for finding the solution by substituting the expression which is already solved for one variable into the other equation.

Note : If the coefficient of one variable is 1 or −1 or if one of the two equations is easily solved for one variable, – for example, (x = expression of y) or (y = expression of x) − this substitution method is the best way to find the solution.

Example 1

$$\begin{cases} 2x + y = 3 & \cdots\cdots ① \\ 3x + 2y = 5 & \cdots\cdots ② \end{cases}$$

Step 1. Solve equation ① for the variable y ; $y = -2x + 3$

Step 2. Substitute $y = -2x + 3$ into the other equation ②.

$$3x + 2(-2x + 3) = 5 \; ; \; -x = -1 \; ; \; x = 1$$

Step 3. Substitute $x = 1$ into equation ① ; $2 \cdot 1 + y = 3$; $y = 1$

Step 4. Find the solution.

$$(x, y) = (1, 1)$$

Note : If the coefficients of variables are unknown and the solution for the system of equations is given, then do the following

Step 1. Substitute the solution into the given system of equations.

Step 2. Find the coefficients of variables using the elimination method or substitution method.

Example 2

$$\begin{cases} ax + by = 3 & \cdots\cdots ① \\ 2bx - ay = -4 & \cdots\cdots ② \end{cases}$$

The solution for this system of equations is $(1, -2)$. Find the value of $a + b$.

Step 1. $\begin{cases} a - 2b = 3 & \cdots\cdots ③, \text{ from } ① \\ 2b + 2a = -4 & \cdots\cdots ④, \text{ from } ② \end{cases}$

Step 2.
$$a - 2b = 3$$
$$+) \; \underline{2a + 2b = -4}$$
$$3a \quad\quad = -1 \; ; \; a = -\frac{1}{3}$$

Substituting $a = -\frac{1}{3}$ into ③, $b = -\frac{5}{3}$

Therefore, $a + b = -\frac{1}{3} - \frac{5}{3} = -2$

3. Solving Special Systems

(1) If the coefficients of variables are fractions or decimals
⇒ Change the coefficients to integers. Then solve as usual.

Example

Solve a system $\begin{cases} \frac{1}{2}x - \frac{1}{3}y = 1 & \cdots\cdots ① \\ 0.3x + 0.2y = 0.2 & \cdots\cdots ② \end{cases}$

Step 1. $\begin{cases} 3x - 2y = 6 & \cdots\cdots ① \times 6 \\ 3x + 2y = 2 & \cdots\cdots ② \times 10 \end{cases}$

Step 2.

$$3x - 2y = 6 \cdots\cdots ③$$
$$-) \underline{3x + 2y = 2 \cdots\cdots ④}$$
$$-4y = 4 \quad ; y = -1$$

Substituting $y = -1$ into ③, $x = \frac{4}{3}$

Therefore, $(x, y) = \left(\frac{4}{3}, -1\right)$

(2) $\begin{cases} px = qy + r \\ my = nx + t \end{cases}$ ⇒ Rearrange or simplify in the form of $ax + by = c$

Example

$\begin{cases} 2x = 3y + 5 \\ -y = 4x - 3 \end{cases}$ $\xrightarrow{\text{rearrange}}$ $\begin{cases} 2x - 3y = 5 & \cdots\cdots ① \\ 4x + y = 3 & \cdots\cdots ② \end{cases}$

$\begin{cases} 4x - 6y = 10 & \cdots\cdots ① \times 2 \\ 4x + y = 3 & \cdots\cdots ② \end{cases}$

$$4x - 6y = 10 \cdots\cdots ③$$
$$-) \underline{4x + y = 3 \cdots\cdots ④}$$
$$-7y = 7 \quad ; y = -1$$

Substituting $y = -1$ into ④, $x = 1$

Therefore, $(x, y) = (1, -1)$

$$(3) \begin{cases} \dfrac{a}{x} + \dfrac{b}{y} = c \\ \dfrac{m}{x} + \dfrac{n}{y} = t \end{cases} \Rightarrow \text{Let } \dfrac{1}{x} = A, \ \dfrac{1}{y} = B$$

Example

$$\begin{cases} \dfrac{2}{x} + \dfrac{3}{y} = 1 \\ \dfrac{1}{x} + \dfrac{2}{y} = 2 \end{cases} \xrightarrow{\text{replace}} \begin{cases} 2A + 3B = 1 & \cdots\cdots \text{①} \\ A + 2B = 2 & \cdots\cdots \text{②} \end{cases}$$

$$\begin{cases} 2A + 3B = 1 & \cdots\cdots \text{①} \\ 2A + 4B = 4 & \cdots\cdots \text{②} \times 2 \end{cases}$$

$$\begin{array}{r} 2A + 3B = 1 \quad \cdots\cdots \text{①} \\ -) \ \underline{2A + 4B = 4} \quad \cdots\cdots \text{③} \\ -B = -3 \ \ ; \ B = 3 \ ; \ \dfrac{1}{y} = 3 \ ; \ y = \dfrac{1}{3} \end{array}$$

Substituting $B = 3$ into ②, $A = -4$; $\dfrac{1}{x} = -4$; $x = -\dfrac{1}{4}$

Therefore, $(x, y) = \left(-\dfrac{1}{4}, \dfrac{1}{3}\right)$

---

(4) A system of equations in form of the $A = B = C$

$\Rightarrow$ Rewrite the equations as a system

$$\begin{cases} A = B \\ B = C \end{cases} \text{ or } \begin{cases} A = B \\ A = C \end{cases} \text{ or } \begin{cases} A = C \\ B = C \end{cases}$$

Note : If $A = B = k$, k is a constant, rewrite the equation as a system $\begin{cases} A = k \\ B = k \end{cases}$.

Example

Solve the system $2x - y = x + 3y = y + 5$.

Step 1. (Rewrite)

$$\begin{cases} 2x - y = x + 3y & \cdots\cdots \text{①} \\ 2x - y = y + 5 & \cdots\cdots \text{②} \end{cases}$$

Step 2. (Rearranging ① and ②)

$$\begin{cases} x - 4y = 0 & \cdots\cdots \text{③} \\ 2x - 2y = 5 & \cdots\cdots \text{④} \end{cases}$$

Step 3. (Solve)

$$2x - 8y = 0 \ \cdots\cdots ③ \times 2$$

$$-) \ \underline{2x - 2y = 5 \ \cdots\cdots ④}$$

$$-6y = -5 \quad ; \ y = \frac{5}{6}$$

Substituting $y = \frac{5}{6}$ into ③, $x = \frac{10}{3}$

Therefore, $(x, y) = \left(\frac{10}{3}, \frac{5}{6} \right)$

---

(5) A system of three equations with three variables

⇒ Step 1. Eliminate one of the variables to create a system of equations with two variables.

Step 2. Solve the new system for two variables.

Step 3. Substitute the two values to find the eliminated variable.

Step 4. Find the solution.

Example

Find the solution (x, y, z) for the system $\begin{cases} 2x - y + z = 30 \ \cdots\cdots ① \\ x + 2y + z = 2 \quad \cdots\cdots ② \\ 3x + 2y - z = -18 \cdots\cdots ③ \end{cases}$

Step 1. (To remove z)

$\begin{cases} 5x + y = 12 \quad \cdots\cdots ④ \ (\text{using } ① + ③) \\ 4x + 4y = -16 \ \cdots\cdots ⑤ \ (\text{using } ② + ③) \end{cases}$

To solve equations,

the number of variables

= the number of equations

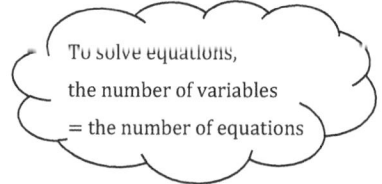

Step 2. (To find (x, y))

$$20x + 4y = 48 \ (\text{using } ④ \times 4)$$

$$-) \ \underline{4x + 4y = -16}$$

$$16x \quad\quad = 64 \quad ; \ x = 4$$

Substituting $x = 4$ into ④, $y = 12 - 20 = -8$

So, $(x, y) = (4, -8)$

Step 3. (To find z)

Substituting $(x, y) = (4, -8)$ into ①,

$$z = 30 - 2x + y = 30 - 2 \cdot 4 - 8 = 14$$

Step 4. $(x, y, z) = (4, -8, \ 14)$

Note :

$$\begin{cases} ax + by = m & \cdots\cdots \text{①} \\ by + cz = n & \cdots\cdots \text{②} \\ cz + ax = t & \cdots\cdots \text{③} \end{cases} \xrightarrow{\text{add all three equations}} 2(ax + by + cz) = m + n + t$$

So, $ax + by + cz = \frac{1}{2}(m + n + t)$

Using ①, $cz = \frac{1}{2}(m + n + t) - m$

Using ②, $ax = \frac{1}{2}(m + n + t) - n$

Using ③, $by = \frac{1}{2}(m + n + t) - t$

At this point, it's easy to find the solution (x, y, z).

---

(6) If a system $\begin{cases} ax + by = c \\ mx + ny = t \end{cases}$ has the special condition $\dfrac{a}{m} = \dfrac{b}{n} = \dfrac{c}{t}$, then

the system has an unlimited number of solutions. (Consistent and dependent)

If the two equations are the same (in that they have the same ratios), that is, form of $0 = 0$, then it's always true.

Example

$$\begin{cases} 2x + 3y = 5 \\ 4x + 6y = 10 \end{cases}$$

$$\Rightarrow \frac{2}{4} = \frac{3}{6} = \frac{5}{10}$$

∴ The system has unlimited solutions.

If the two equations have the same coefficients for the variables but constants, that is, form of $0 = a$ (a is a non-zero number), then it's impossible.

(7) If a system $\begin{cases} ax + by = c \\ mx + ny = t \end{cases}$ has the special condition $\dfrac{a}{m} = \dfrac{b}{n} \neq \dfrac{c}{t}$, then

The system does not have a solution. (Inconsistent)

$ax + by + c = 0 \xrightarrow{\text{transform}} y = px + q$

(Equation $\xrightarrow{\text{transform}}$ Function)

If $y_1 = px + q_1$ and $y_2 = px + q_2$, $q_1 \neq q_2$, then y_1 and y_2 are parallel.

So, there is no intersection point.

Therefore, $\dfrac{a}{m} = \dfrac{b}{n} \neq \dfrac{c}{t}$ $\Rightarrow$ The system has no solution.

Example

$$\begin{cases} 3x + y = 2 \\ 6x + 2y = 3 \end{cases}$$

$$\Rightarrow \quad \frac{3}{6} = \frac{1}{2} \neq \frac{2}{3}$$

∴ The system has no solution.

4. Steps for Solving Word Problems (with Two Variables)

Step 1. Assign two variables to solve the unknowns.

Step 2. Create a system of equations and solve it, using any applicable methods.

Step 3. Check your answer.

5. Graphing a System of Equations

For any constants $a, b,$ and $c,$ a system $\begin{cases} ax + by = c \\ a'x + b'y = c' \end{cases}$ in a coordinate plane creates one of three types of cases : parallel, intersecting at one point, or coinciding.

To graph a system, find the x-intercept and y-intercept for each equation. Then graph both equations on the same coordinate plane.

The solution of a system of equations will be the intersection point of the two lines.

Note :

x -intercept is the x-coordinate of the point where the line crosses (intersects) the x-axis.
(x-intercept is the value of x when $y = 0$.)

y -intercept is the y-coordinate of the point where the line crosses (intersects) the y-axis.
(y-intercept is the value of x when $x = 0$.)

See Chapter 2 in Algebra, Part III Functions for more information.

Example

Solve the system $\begin{cases} 2x + y = 4 \\ 4x - y = 2 \end{cases}$ by graphing.

$2x + y = 4 \implies y = -2x + 4 \quad \therefore x\text{-intercept}: (2, 0) \text{ and } y\text{-intercept}: (0, 4)$

$4x - y = 2 \implies y = 4x - 2 \quad \therefore x\text{-intercept}: \left(\frac{1}{2}, 0\right) \text{ and } y\text{-intercept}: (0, -2)$

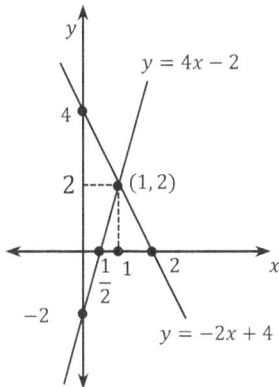

By graphing, we find the intersection point $(x, y) = (1, 2)$.
So, $x = 1, \ y = 2$

Exercises

#1. Solve each system.

(1) $\begin{cases} x + y = 4 \\ 2x - y = 5 \end{cases}$

(2) $\begin{cases} 2x + y = 3 \\ 2x + 3y = 4 \end{cases}$

(3) $\begin{cases} 2x + y = 7 \\ 3x + 2y = 5 \end{cases}$

(4) $\begin{cases} x - y = 5 \\ 2x + 5y = 3 \end{cases}$

(5) $\begin{cases} 3x - y = 0 \\ 5x - 2y = -3 \end{cases}$

(6) $\begin{cases} 4x - 3y = 6 \\ 6x - 2y = -6 \end{cases}$

#2. Find the value of ab for each system.

(1) $\begin{cases} ax - by = -2 \\ bx + 2y = a \end{cases}$ with solution $(3,2)$

(2) $\begin{cases} x + 5y = -3 \\ 2x - by = 5 \end{cases}$ with solution $(a, -1)$

(3) $\begin{cases} -2x + y = 5 \\ x - 2y = -1 \end{cases}$ with solution (a, b)

(4) $\begin{cases} 3x - by = 2 \\ ax + y = -2 \end{cases}$ with solution $(b - 1, 2)$

#3. The system $\begin{cases} 2x - y = 3 \\ x + 3y = 5 \end{cases}$ has a solution $(a + 1, b - 1)$. Find the value of $(a + b)^2 - (a - b)^2$.

#4. The system $\begin{cases} 2x + 3y = 5 \\ -x - 2y = -3 \end{cases}$ has a solution (a, b).

Find the solution for the system $\begin{cases} (3 - a)x + 2y = -2 \\ 2x + 3y = 2b + 1 \end{cases}$.

#5. The solution of the system $\begin{cases} 3x - 2y = -2 \\ (k - 1)x + y = -3 \end{cases}$ is the same as the solution of the equation $2x - y = 3$. Find the constant k.

#6. Two systems $\begin{cases} 2x + by = 4 \\ x + 2y = -3 \end{cases}$ and $\begin{cases} x - 3y = 2 \\ ax + 2y = -1 \end{cases}$ have the same solution.

Find the value of $a + b$.

#7. Solve each system.

(1) $\begin{cases} x = 2y + 1 \\ x - y = 3 \end{cases}$

(2) $\begin{cases} y = x - 3 \\ 2x - y = 2 \end{cases}$

(3) $\begin{cases} \dfrac{1}{2}x + \dfrac{1}{3}y = 2 \\ \dfrac{2}{3}x - \dfrac{3}{4}y = -\dfrac{11}{12} \end{cases}$

(4) $\begin{cases} \dfrac{2}{x} + \dfrac{3}{y} = 4 \\ \dfrac{1}{2x} - \dfrac{1}{y} = 2 \end{cases}$

(5) $\begin{cases} 0.2x - 0.5y = 0.25 \\ -0.3x + 0.4y = -0.2 \end{cases}$

(6) $\begin{cases} 3x - y = -5 \\ 6x - 2y = 3 \end{cases}$

(7) $\begin{cases} x = \dfrac{1}{2}y + 1 \\ 2x - 4y = -4 \end{cases}$

(8) $\begin{cases} -\dfrac{2}{3}x + \dfrac{1}{4}y = -2 \\ 2x - \dfrac{3}{4}y = 6 \end{cases}$

(9) $\begin{cases} \dfrac{2}{x} + \dfrac{2}{y} = 1 \\ \dfrac{1}{x} - \dfrac{1}{y} = -1 \end{cases}$

(10) $\dfrac{x+1}{2} + \dfrac{y-1}{3} = \dfrac{2x-1}{3} + \dfrac{y+2}{4} = 2x + \dfrac{y}{2}$

(11) $\begin{cases} 2x - 3y = -4 \\ -3y + z = 2 \\ z + 2x = -6 \end{cases}$

(12) $\begin{cases} x : y + 1 = 2 : 3 \\ 3 : y - 1 = 4 : x - 1 \end{cases}$

#8. The system $\begin{cases} \frac{a+1}{2}x - \frac{3}{4}y = -2 \\ 5x + \frac{b-1}{2}y = 4 \end{cases}$ has an unlimited number of solutions.

Find the value of $a + b$.

#9. The system $\begin{cases} a(x - y) + \frac{y}{2} = -1 \\ -\frac{x}{2} - \frac{1}{a}y = 3 \end{cases}$ has no solution. Find the value of a.

#10. The system $\begin{cases} 2kx - (3x + y) = 2y \\ -(k - 1)x + 2y = kx \end{cases}$ has a solution other than (0,0).

Find the value of the constant k.

#11. Find the value of $a + b$ for the following systems

(1) $\begin{cases} 2x - \frac{y}{3} = \frac{2}{3} \\ (x - y) : 3 = -1 : 1 \end{cases}$ with solution $(a, b - 1)$.

(2) $\begin{cases} \frac{3}{x} + \frac{2}{y} = b \\ \frac{1}{x} + \frac{2}{y} = \frac{1}{2} \end{cases}$ with solution $(a, 2a)$.

(3) $ax + (b - 1)y = 2ax - 3y + 5 = x + by - 1$ with solution $(2, 3)$.

#12. The system $\begin{cases} 3x - 2y = k \\ -2x + y = 3 \end{cases}$ has the solution (a, b) with the condition $a : b = 1 : 3$

Find the constant k.

#13. Find the value of $x + y$ for variables x and y that satisfy the equations $2^x \cdot 8^y = 32$ and $3^{x+1} \cdot 9^{y-1} = 3^3$

#14. Find the value of $\dfrac{1}{x} - \dfrac{1}{y}$ for variables x and y that satisfy the system

$\begin{cases} 2x - xy - 2y - 3 = 0 \\ 3x + 2xy - 3y + 1 = 0 \end{cases}$.

#15. The perimeter of a rectangle is 18 inches. The length of the rectangle is 3 inches shorter than twice its width. What is the area of the rectangle?

#16. Movie ticket prices are $6 for children and $9 for adults. Nichole pays $84 for 12 people. How many children are in her group?

#17. Apples and peaches are mixed in a box. There are 3 less apples than three times the number of peaches. Two times the total number of apples and peaches is 10. How many apples and peaches are in the box?

#18. Richard prepares a bag of candies for kids. If each kid gets 8 candies, then 8 candies will be left. If they each get 10 candies then Richard will be short 6 candies. How many candies are in Richard's bag?

#19. Nichole has quarters and dimes worth $2.55 in her purse. The number of dimes is two less than three times the number of quarters. How many quarters and dimes are in her purse?

#20. If you add the ten's digit and the one's digit of a certain two-digit integer, the sum is 12. If the digits of the number are interchanged, the new number will be 12 less than twice the original number. Find the original number.

#21 If 30 ounces of salt solution containing a x% of salt solution is added to 40 ounces of salt solution containing a y% of salt solution, it produces a salt solution that is 15% salt. If 30 ounces of a salt solution containing a y% of salt solution is added to 40 ounces of a salt solution containing a x% of salt solution, it produces a salt solution that is 18% salt. Find the values of x and y.

#22 . Richard wants to produce 70 ounces of salt solution that is 8% salt by adding water after mixing salt solution that is 5% salt with salt solution that is 10% salt. The amount of a 10% of salt solution is three times as much as the amount of a 5% of salt solution. How much water should he add?

#24. Nichole started to run at 9:50AM at a speed of 8 miles per hour and then walked the rest of the way at 3 miles per hour. She arrived at 10:40AM. If Nichole went to the park which was 4 miles away, how many miles did she run?

#23. Nichole wants to produce 29 liters of 20% alcohol solution by adding alcohol after mixing two alcohol solutions that are 4% alcohol and 3% alcohol separately. The amount of alcohol solution that is 3% alcohol is twice the amount of alcohol solution that is 4% alcohol. How many liters of alcohol must be added?

#25. Richard starts a trail ride in a parking lot. He rides up a long hill on *A* trail at 4 miles per hour and comes down the hill on *B* trail going 12 miles per hour. His ride takes 1 hour 20 minutes total. The total distance of *A* trail and *B* trail is 10 miles. How many miles long is *B* trail?

#26. Richard and Nichole want to finish a job. Richard works alone for 3 hours and leaves. Nichole comes and works alone the rest of the job for another 3 hours, thereby finishing the job. OR if Richard works alone for 6 hours and then Nichole works alone for 2 hours for the rest of the job after Richard is done, the job is also completed. How long will it take Richard to finish the job by himself the entire time?

#27. 5 years ago, Nichole was 5 years less than one-third her mom's age. In 6 years, her mom will be 10 years more than twice Nichole's age at that time. How old was Nichole's mom when Nichole was 15?

#28. Richard walks from home to a library at 3 miles per hour. 20 minutes after Richard leaves, Nichole rides a bike at 8 miles per hour from home to the library. They arrive at the same time. How long does it take Richard to meet Nichole at the library?

#29. If Richard drives a car from home to the doctor's office at 50 miles per hour, he will arrive at the office 5 minutes earlier than his appointment time. If he drives a car at 40 miles per hour on the same route, he will arrive 10 minutes late to his appointment. How far is the office from Richard's home?

#30. Richard and Nichole jog towards each other from two opposite starting points 1 mile apart. Nichole jogs 1.5 times faster than Richard. How fast does Nichole have to jog if they want to meet each other in 30 minutes?

#31. There were 44 boys and girls in a math club last summer. This year, 25% of the boys quit and 15% of the girls joined the club again. Now the club has 41 members. How many boys and girls are in the club now?

#32. Six years ago, Nichole was three times as old as Richard. Four years from now, Nichole will be twice as old as Richard. How old is Richard now?

#33. Solve each system by graphing.

(1) $\begin{cases} x + y = 4 \\ 2x - y = 5 \end{cases}$

(2) $\begin{cases} 2x - y = 5 \\ 4x - 2y = 6 \end{cases}$

(3) $\begin{cases} 3x + 4y = 5 \\ 6x + 8y = 10 \end{cases}$

Algebra

**Part II
Expressions**

Chapter 5

**Systems of Inequalites
Level II**

CHAPTER
5

Chapter 5   Systems of Inequalities

5-1 Systems of Inequalities

    1.   Definition

    2.   Solving Systems of Inequalities

    3.   Solving Special Systems

5-2   Graphs of Linear Inequalities with Two Variables

    1.   Half-planes

    2.   Graphing Linear Inequalities

    3.   Graphing Systems of Inequalities

Chapter 5. Systems of Inequalities

5-1 Systems of Inequalities

1. Definition

A *system of inequalities* is a pair of two or more linear inequalities.

The *solution* for a system of inequalities is the value or range which satisfies both inequalities at the same time.

2. Solving Systems of Inequalities

Step 1. Solve each inequality and find the solutions for each.

Step 2. Graph the solutions on a number line.

Step 3. Find the common range or value.

When $a < b$,

(1) $\begin{cases} x < a \\ x < b \end{cases} \Rightarrow x < a$

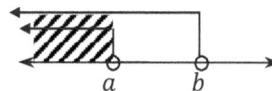

(2) $\begin{cases} x > a \\ x > b \end{cases} \to x > b$

(3) $\begin{cases} x > a \\ x < b \end{cases} \Rightarrow a < x < b$

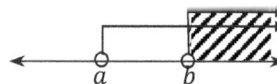

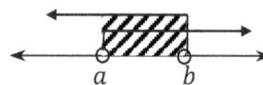

3. Solving Special Systems

(1) Solve parentheses by using the distributive property.

(2) Change fraction or decimal coefficients of variables to integers.

(3) Rewrite the system $A < B < C$ as $\begin{cases} A < B \\ B < C \end{cases}$.

$A = B = C \Rightarrow$

$\begin{cases} A = B \\ B = C \end{cases}$ or $\begin{cases} A = B \\ A = C \end{cases}$ or $\begin{cases} A = C \\ B = C \end{cases}$

But, $A < B < C \Rightarrow \begin{cases} A < B \\ B < C \end{cases}$ only

$\begin{cases} A < B \\ A < C \end{cases}$ ($\because$ we can't compare B and C)

$\begin{cases} A < C \\ B < C \end{cases}$ ($\because$ we can't compare A and B)

Don't rewrite the system when only B has an unknown variable.

For example,

$2 < 2x < 3 \Rightarrow 1 < x < \frac{3}{2}$

constant — x term — constant

(4) $\begin{cases} x \le a \\ x \ge a \end{cases}$ $\Rightarrow$ $x = a$ (This is the only solution.)

(5) If there is no common value or range,

$\begin{cases} x < a \\ x \ge a \end{cases}$ $\begin{cases} x < a \\ x > a \end{cases}$ $\begin{cases} x < a \\ x \ge b \end{cases}$, $a < b$

$\Rightarrow$ no solution $\Rightarrow$ no solution $\Rightarrow$ no solution

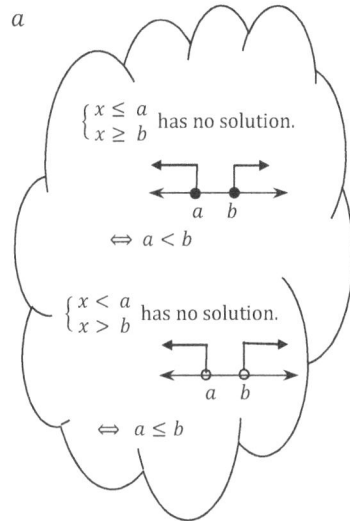

$\begin{cases} x \le a \\ x \ge b \end{cases}$ has no solution.

$\Leftrightarrow a < b$

$\begin{cases} x < a \\ x > b \end{cases}$ has no solution.

$\Leftrightarrow a \le b$

(6) Conditions for having solutions :

1) $\begin{cases} x \le a \\ x \ge b \end{cases}$ $\Rightarrow b \le a$

($\because$ if $b = a$, then $\begin{cases} x \le a \\ x \ge a \end{cases}$ $\therefore x = a$ is the solution.)

2) $\begin{cases} x \le a \\ x > b \end{cases}$ $\Rightarrow b < a$

($\because$ if $b = a$, then $\begin{cases} x \le a \\ x > a \end{cases}$ $\therefore$ There is no solution. Therefore, $b \ne a$ $(b < a)$.)

3) $\begin{cases} x < a \\ x \ge b \end{cases}$ $\Rightarrow b < a$

($\because$ if $b = a$, then $\begin{cases} x < a \\ x \ge a \end{cases}$ $\therefore$ There is no solution. Therefore, $b \ne a$ $(b < a)$.)

4) $\begin{cases} x < a \\ x > b \end{cases}$ $\Rightarrow b < a$

($\because$ if $b = a$, then $\begin{cases} x < a \\ x > a \end{cases}$ $\therefore$ There is no solution. Therefore, $b \ne a$ $(b < a)$.)

(7) For any $a > 0$,

$\begin{cases} |x| \le a \Rightarrow -a \le x \le a \\ |x| > a \Rightarrow x > a \text{ or } x < -a \end{cases}$

$|x| \le a$ $|x| > a$

5-2 Graphs of Linear Inequalities with Two Variables

1. Half-planes

The graph of a linear inequality with two variables (x and y) consists of points in the coordinate plane whose coordinate (x, y) makes the linear inequality true.

For a linear inequality $ax + by \geq c$ or $ax + by \leq c$,
the coordinates on the line $ax + by = c$ are solutions to the linear inequality, so we use a straight line.

For a linear inequality $ax + by > c$ or $ax + by < c$,
the coordinates on the line $ax + by = c$ are not solutions to the linear inequality, so we use a dotted line instead of a straight line.

The straight line or dotted line separates the coordinate plane into two half-planes.

The inequality symbols " $\geq$ " or " $\leq$ " represent closed half-planes and the inequality symbols " $>$ " or " $<$ " represent open half-planes.

2. Graphing Linear Inequalities

Each half-plane region is the solution to the inequality.
To find the solution, choose a point in one half-plane and substitute it into the inequality.
If the point makes the inequality true, then the region including the point is the solution to the inequality. Shade the solution region.

Example 1

Graph the linear inequality $2x + y < 4$.

Consider the graph of the line $y = -2x + 4$.

The line $y = -2x + 4$ has slope -2 and y-intercept $(0, 4)$.

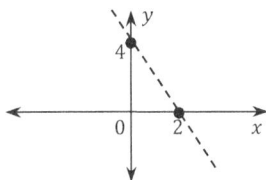

Choose a point $(-1, 0)$ in one half-plane and substitute the point into the inequality.

Since $2x + y = 2 \cdot (-1) + 0 = -2$ and $-2 < 4$, the inequality is true. So the region including the point $(-1, 0)$ is the solution to $-2x + y < 4$.

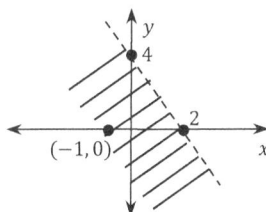

Example 2

Graph the linear inequality $2x - 3y \geq 6$.

$2x - 3y \geq 6 \implies 3y \leq 2x - 6 \implies y \leq \frac{2}{3}x - 2$

Graph the line $y = \frac{2}{3}x - 2$.

The line $y = \frac{2}{3}x - 2$ has x-intercept $(3, 0)$ and y-intercept $(0, -2)$.

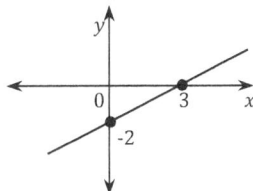

Choose a point $(0, 0)$ as a testing point.

Substituting $x = 0$ and $y = 0$ into the inequality, we get $2x - 3y = 2 \cdot 0 - 3 \cdot 0 = 0$.

Since $0 \geq 6$ is not true, the region including the point $(0, 0)$ is not the solution. Shade the region which does not include the point $(0, 0)$.

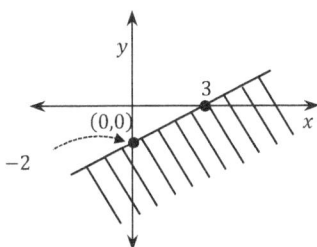

3. Graphing Systems of Inequalities

To solve a system of inequalities, find the solution region for each inequality. Graph all inequalities on one coordinate plane. The overlapping region of the graph is the solution to the system of inequalities.

Every point in the overlap region makes all the inequalities true.

Example

Graph a system of linear inequalities $\begin{cases} 2x - y \geq 1 \\ x - 3y > 4 \end{cases}$.

Since $2x - y \geq 1 \Rightarrow y \leq 2x - 1$, graph the line $y = 2x - 1$.

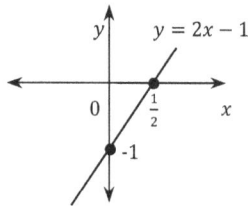

Choose a testing point $(0, 0)$.

Substituting $x = 0$ and $y = 0$ into the inequality,

$2x - y = 2 \cdot 0 - 0 = 0 \geq 1$: This is not true.

So,

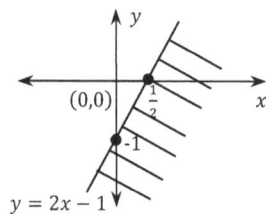

The shaded region is the solution to the linear inequality $2x - y \geq 1$.

Since $x - 3y > 4 \Rightarrow y < \frac{1}{3}x - \frac{4}{3}$, graph the line $y = \frac{1}{3}x - \frac{4}{3}$.

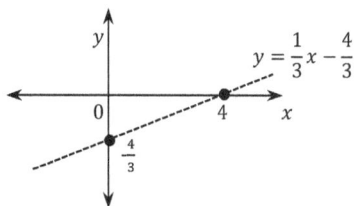

Choose a testing point $(0, 0)$.

Substituting $x = 0$ and $y = 0$ into the inequality,

$x - 3y = 0 - 3 \cdot 0 = 0 > 4$: This is not true.

So,

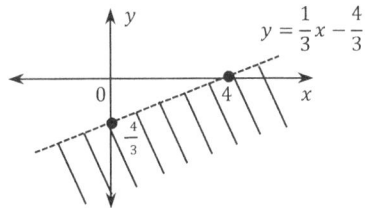

Graph both inequalities on the same coordinate plane. Shade the overlapping solution region more darkly than the other solution regions.

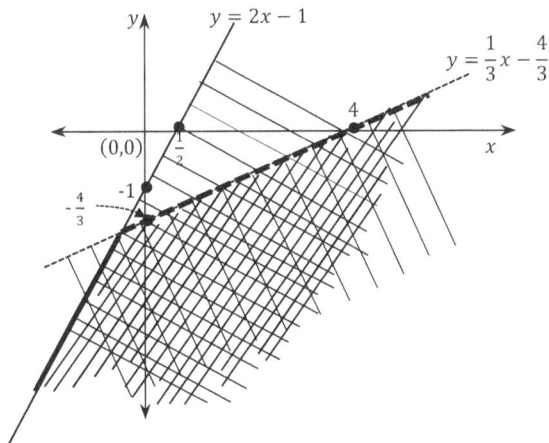

Exercises

#1. Find the range of x for the following systems

(1) $\begin{cases} x \geq -2 \\ x < 4 \end{cases}$ (2) $\begin{cases} x > -3 \\ x > 0 \end{cases}$ (3) $\begin{cases} x < -1 \\ x \leq 2 \end{cases}$

(4) $\begin{cases} x \leq 1 \\ x \geq 3 \end{cases}$ (5) $\begin{cases} x \geq 2 \\ x < 2 \end{cases}$ (6) $\begin{cases} x \leq -3 \\ x \geq -3 \end{cases}$

#2. Solve the following systems

(1) $\begin{cases} 3x - 1 < 4x + 2 \\ x + 3 \geq -2x \end{cases}$ (2) $\begin{cases} -2x + 4 > 3x - 6 \\ 3x - 5 \geq x + 3 \end{cases}$

(3) $\begin{cases} x - 2 > -5 \\ 5x - 3 < 2x + 3 \end{cases}$ (4) $\begin{cases} 4x - 3 > 5x - 2 \\ 2x - 2 > 3x - 5 \end{cases}$

(5) $\begin{cases} \frac{x+2}{2} \leq \frac{x-2}{3} + x \\ 2(x - 2) - \frac{x}{2} > 2x - 6 \end{cases}$ (6) $x - 3 < \frac{x}{2} + 1 \leq 3x - 1$

(7) $0.5x + \frac{x}{2} + 1 > 0.7x - 0.2 > 0.3x - 0.6$ (8) $\begin{cases} 2.1x > 3.6 - 0.9x \\ 3(x - 1) - 2 < \frac{1}{2}(3x + 1) \end{cases}$

#3. Find the range of $y = -2x + 3$ when x satisfies the following systems

(1) $\begin{cases} x - 3 \leq 2 \\ -3x + 2 < 4 \end{cases}$

(2) $\begin{cases} 2x - 3 < 4x - 1 \\ 5x - 2 \geq 3x + 2 \end{cases}$

(3) $\frac{x-1}{3} \leq \frac{1}{4}(x - 3) < \frac{5-3x}{4}$

(4) $\begin{cases} 2.1x > 3.6 - 0.9x \\ 3(x - 1) - 2 < \frac{1}{2}(3x + 1) \end{cases}$

#4. Find the value of k for the following conditions

(1) The system $\begin{cases} x + 5 < 2k \\ 3x - 2 \geq 4 \end{cases}$ has the solution $2 \leq x < 5$.

(2) The system $\begin{cases} \frac{2x+1}{3} > \frac{x-3}{5} \\ 0.6x - 2.4 < kx - 0.8 \end{cases}$ has the solution $-2 < x < 2$.

(3) The system $\begin{cases} -x + 2 \leq 0 \\ \frac{x}{2} + 3 \leq -k + 5 \end{cases}$ has the solution $x = 2$.

(4) The system $\begin{cases} \frac{k-x}{2} \leq x + 5 \\ 3 - 2x < 3x - 2 \end{cases}$ has the solution $x \geq 3$.

(5) The system $\begin{cases} 2x + 3 \leq 4x - 5 \\ 3(x - k) \leq x + 3 \end{cases}$ has only one solution.

(6) The system $3 < \dfrac{k-4x}{-2} < 5$ has the solution $1 < x < 2$.

#5. Find the range of k for the following conditions

(1) The system $\begin{cases} 2x \leq 5 - k \\ 3x - 3 \geq 2x - 1 \end{cases}$ has no solution.

(2) The system $\begin{cases} x - 3 \leq 2x - 6 \\ 5x + k < 3x + 1 \end{cases}$ has no solution.

(3) The system $\begin{cases} 2x + 3 \leq -5 \\ x + k > 1 \end{cases}$ has only one integer in the solution.

(4) The system $\begin{cases} x - 3 \geq 0 \\ 3x + k \leq 2x + 3 \end{cases}$ has solutions.

#6. Find the sum of all integers that satisfy the following systems

(1) $\begin{cases} 3x - 5 \le 7 \\ \frac{x-1}{2} < x + 3 \\ 2x - 5 < 5x + 4 \end{cases}$

(2) $\begin{cases} |x| \le 5 \\ |x| > 2 \end{cases}$

#7. The system $\begin{cases} x - 1 \ge 2x - 4 \\ \frac{x+k}{2} < 3x - 2 \end{cases}$ has 5 integers in the solution. What is the minimum value for k?

#8. 3 more than twice a number is less than or equal to 8, and -1 is less than one fourth of the number.

(1) Find the range of the number.

(2) Find the sum of the maximum integer and minimum integer that satisfies the system.

(3) Solve the system with the condition, -1 is greater than one fourth of the number, instead of the second inequality in the system shown in (1). Find the maximum integer that satisfies the new system.

#9. The sum of three consecutive positive integers is greater than 60 and less than 65.

Find the largest number.

#10. The lengths of three sides of a triangle are $x - 4$, $x + 1$, and $x + 3$. Find the range of x.

#11. Nichole wants to produce new salt solution by adding salt into 20 ounces of a 15% salt solution. It will be a salt concentration greater than that of a 20% salt solution and less than that of a 25% salt solution. How much salt does she need to add?

#12. Richard jogs 12 miles. He begins by walking at a speed of 5 miles per hour. He then runs at a speed of 10 miles per hour for the remaining distance. If he wants to take at least 1 hour 45 minutes and at most 2 hours to complete his route, what is the longest distance he should walk?

#13. Nichole wants to reorganize all the books in her bookshelf. If she puts 30 books in each shelf, then 5 books will be left over. If she puts 35 books in each shelf, then there will be at least 20 books, but less than 25 books on the last shelf. How many shelves are in the bookshelf?

#14. Solve the following inequalities

(1) $|x - 3| < 4$

(2) $|x + 2| < 3x - 4$

(3) $|x + 2| + |3 - x| > 10$

#15 Graph the following systems of linear inequalities

(1) $\begin{cases} y < x \\ y \geq -x \end{cases}$

(2) $\begin{cases} 2x + y \leq 4 \\ x - y \geq 2 \end{cases}$

(3) $\begin{cases} 3x + y \leq 6 \\ 2x + 3y > 4 \end{cases}$

(4) $\begin{cases} y \leq 2 \\ x + y > 3 \\ x > -4 \end{cases}$

Algebra

Part II
Expressions

Chapter 6

Factorization
Level III

CHAPTER 6

Chapter 6 Factorization

Chapter 6. Factorization

6-1 Factoring Polynomials

1. Definition

(1) The Greatest Common Factor (GCF)

The Greatest Common Factor (GCF) is the product of their common factors raised to their lowest powers.

Example

$$90x^2y^3 + 12x^4y^5$$

It cannot be factored further.

Since $90 = 2 \cdot 3^2 \cdot 5$ (prime factorization) and $12 = 2^2 \cdot 3$ (prime factorization),

the common factors are 2, 3, x, and y.

$$\therefore \text{ GCF} = 2^1 \cdot 3^1 \cdot x^2 \cdot y^3 = 6x^2y^3$$

(2) Factorization

Factorization is an expression of a polynomial as a product of its greatest common factor and two or more prime polynomials.

Example

$$3a^2 + 15a = 3a\,(a + 5)$$

Remaining factor when the GCF is factored out of the polynomial.

GCF of $3a^2$ and $15a$

Example

$$2a(x - 3) - 4(x - 3) = 2\,(x - 3)(a - 2)$$

GCF of $2a(x - 3)$ and $4(x - 3)$

Note :

$(x + 2)(x + 3) = x^2 + 5x + 6$: *Multiplying two polynomials using the distributive property (Expanding)*

$x^2 + 5x + 6 = (x + 2)(x + 3)$: *Factoring polynomials as a product of two prime polynomials*

6-2 Factorization Formulas

Note :

- *Quadratic : The highest power in the polynomial is 2.*
- *Trinomial : A polynomial containing 3 terms*
- *Quadratic trinomial is the form of $ax^2 + bx + c$, $a \neq 0$*

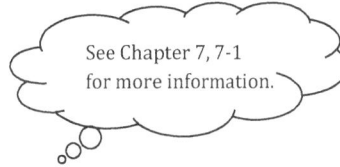

See Chapter 7, 7-1 for more information.

1. The Form of a Perfect Square

$$a^2 + 2ab + b^2 = (a + b)^2$$

$$a^2 - 2ab + b^2 = (a - b)^2$$

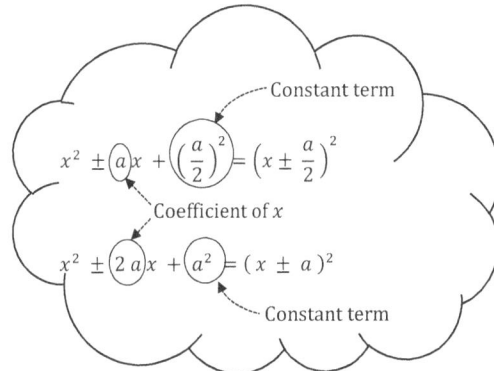

Constant term

$$x^2 \pm a\,x + \left(\frac{a}{2}\right)^2 = \left(x \pm \frac{a}{2}\right)^2$$

Coefficient of x

$$x^2 \pm 2\,a\,x + a^2 = (x \pm a)^2$$

Constant term

Example

$$x^2 + 6x + 9 = (x + 3)^2$$

$2 \cdot 3 \qquad 3^2$

$$x^2 + 5x + 6 \neq (\cdots)^2$$

$2 \cdot \dfrac{5}{2} \qquad 6 \neq (\dfrac{5}{2})^2$

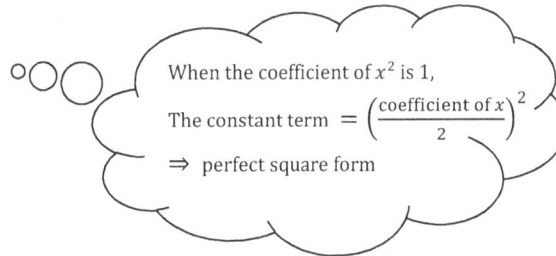

When the coefficient of x^2 is 1,

The constant term $= \left(\dfrac{\text{coefficient of } x}{2}\right)^2$

$\Rightarrow$ perfect square form

Note :

- $a - b = -(b - a)$
- $(a - b)^2 = (-(b - a))^2 = (b - a)^2$
- $(a - b)^3 = (-(b - a))^3 = -(b - a)^3$
- $(a - b)^4 = (-(b - a))^4 = (b - a)^4$

2. The Form of $(a+b)(a-b)$

$$a^2 - b^2 = (a+b)(a-b)$$

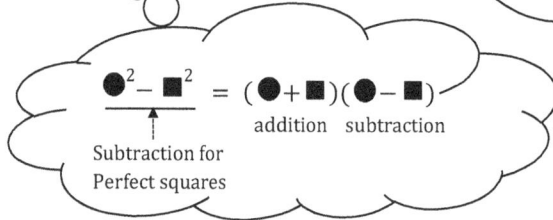

$a^2 + b^2 \neq (\cdots)(\cdots)$

But $a^3 + b^3 = (\cdots)(\cdots)$

$a^3 + b^3 = (a+b)(a^2 - ab + b^2)$

Perfect cubes

$a^3 - b^3 = (a-b)(a^2 + ab + b^2)$

$$\dfrac{\bullet^2 - \blacksquare^2}{} = (\bullet + \blacksquare)(\bullet - \blacksquare)$$

Subtraction for Perfect squares

addition subtraction

Note :

$$-a^2 + b^2 = b^2 - a^2 = (b-a)(b+a)$$

$4x^2 = (2x)(2x) = (2x)^2$: perfect square

$9 = 3 \cdot 3 = 3^2$: perfect square

Example

$$4x^2 - 9 = (2x)^2 - (3)^2 = (2x+3)(2x-3)$$

3. Quadratic Trinomial

$x^2 + (a+b)x + \underline{ab} = (x+a)(x+b)$

① If the constant term $> 0 \Rightarrow \begin{cases} (a > 0 \text{ and } b > 0) \\ \text{or} \\ (a < 0 \text{ and } b < 0) \end{cases}$

② If the constant term $< 0 \Rightarrow \begin{cases} (a > 0 \text{ and } b < 0) \\ \text{or} \\ (a < 0 \text{ and } b > 0) \end{cases}$

(1) If The Leading Coefficient is 1,

$$x^2 + (a+b)x + ab = (x+a)(x+b)$$

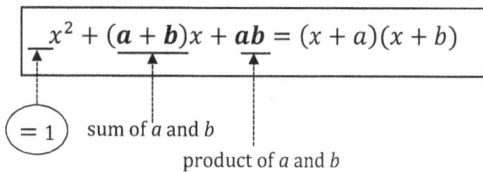

= 1

sum of a and b

product of a and b

① Step 1 Find all the possible numbers which factor the constant ab.

② Step 2 Check to see if the sum of the two numbers found in Step 1 is the same as the coefficient of x.

③ Step 3 Factor $(x+a)(x+b)$.

Example 1

$$x^2 + 3x + 2 = (x + 1)(x + 2)$$

$1 + 2 \qquad 1 \times 2$

$$2 = \begin{cases} 1 \times 2 & \Rightarrow 1 + 2 = 3 & ; \text{ the coefficient of } x \\ (-1)(-2) & \Rightarrow (-1) + (-2) = -3 & ; (\times) \text{ wrong} \end{cases}$$

product of two numbers

sum of two numbers

Example 2

$$x^2 - 3x - 4 = (x - 4)(x + 1)$$

$-4 + 1 \qquad -4 \times 1$

$$-4 = \begin{cases} (-1) \times 4 & \Rightarrow (-1) + 4 = 3 & ; (\times) \text{ wrong} \\ 1 \times (-4) & \Rightarrow 1 + (-4) = -3 & ; \text{ the coefficient of } x \\ 2 \times (-2) & \Rightarrow 2 + (-2) = 0 & ; (\times) \text{ wrong} \end{cases}$$

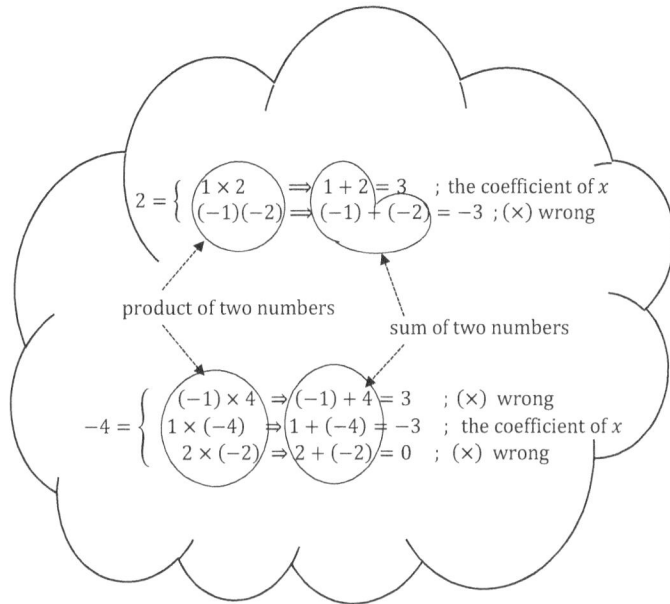

(2) If The Leading Coefficient is not 1,

$$acx^2 + (ad + bc)x + bd = (ax + b)(cx + d)$$

$\neq 1$

Note :

$$acx^2 + (ad + bc)x + bd = (ax + b)(cx + d)$$

Step1
(factor)

Step2
(product)

Step1
(factor)

a b $\rightarrow bc$
c d $\rightarrow ad$ $\Longrightarrow bc + ad$

Step3
(add)

Step4
(check)

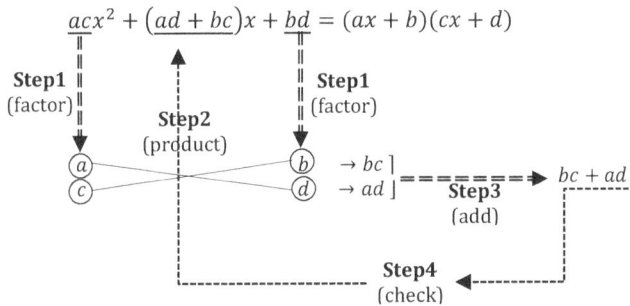

①   Step 1   Find all the possible numbers which factor the coefficient of x^2 and the constant

②   Step 2   Product the numbers diagonally

③   Step 3   Add the numbers obtained from step 2

④   Step 4   Check to see if the sum obtained from Step 3 is the same as the coefficient of x

⑤   Step 5   Factor as $(ax + b)(cx + d)$

Example 1

$$2x^2 + 7x + 6 = (x + 2)(2x + 3)$$

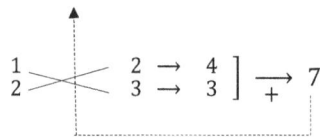

$2 = \mathbf{1} \times \mathbf{2}$ or 2×1
or $(-1) \times (-2)$
or $(-2) \times (-1)$
$6 = 1 \times 6$ or 6×1
or $\mathbf{2} \times \mathbf{3}$ or 3×2
or $(-1) \times (-6)$ or $(-6) \times (-1)$
or $(-2) \times (-3)$ or $(-3) \times (-2)$

Example 2

$$3x^2 + 2x - 5 = (x - 1)(3x + 5)$$

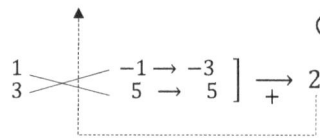

Consider all possibilities
$3x^2 + ②x - 5 = (x - 1)(3x + 5)$

Example 3

$$9x^2 - 12x + 4 = (3x - 2)(3x - 2) = (3x - 2)^2$$

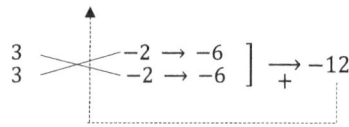

$$x^2 + \mathbf{6}x + \mathbf{9} = (x + 3)^2$$

$2 \cdot 3 \qquad 3^2$: type of
$x^2 + 2ax + a^2$
$= (x + a)^2$

$$x^2 + \mathbf{6}x + \mathbf{9} = (x + 3)(x + 3) = (x + 3)^2$$

A perfect square type of factorization
is a special case for factoring
quadratic trinomials.

6-3 More Factorization

1. If each term has common factors,

⇒ Find the GCF first and express the polynomial as a product of the GCF and prime polynomials.

Note :

$$ax^2 + 4ax + 4a$$

$$\xrightarrow[\text{find the GCF}]{} \quad a(x^2 + 4x + 4)$$

$$\xrightarrow[\text{factor}]{} \quad a(x + 2)^2$$

Example

$$2x^2 + 6x + 4$$

$$
\begin{array}{l}
1 \quad\quad 2 \to 4 \\
2 \quad\quad 2 \to 2
\end{array} \Big] \xrightarrow[+]{} 6
$$

$$= (x + 2)(2x + 2)$$

$$= 2(x + 2)(x + 1)$$

Begin by removing a common factor, and then factoring further.

That is, $2x^2 + 6x + 4 = 2(x^2 + 3x + 2)$

$$x^2 + 3x + 2$$

$$
\begin{array}{l}
1 \quad\quad 1 \to 1 \\
1 \quad\quad 2 \to 2
\end{array} \Big] \xrightarrow[+]{} 3
$$

$$= (x + 1)(x + 2)$$

$$\therefore\ 2x^2 + 6x + 4 = 2(x^2 + 3x + 2) = 2(x + 1)(x + 2)$$

2. If a polynomial contains common terms,

⇒ Substitute the common terms with other variables. Then factor the resulting polynomial.

Rewrite it with common terms.

Example

$(a - 1)^2 + 2(a - 1) - 3 = A^2 + 2A - 3$ substitute $a - 1$ as A

$$= (A + 3)(A - 1) \quad \text{factor}$$

$$= (a - 1 + 3)(a - 1 - 1) \quad \text{replace } A \text{ with } a - 1$$

$$= (a + 2)(a - 2)$$

3. If polynomials containing more than 3 terms have no common factors,

⇒ Factor by groups considering the combined terms have a common term.

Example

$$x^3 + 3x^2 - 4x - 12 = (x^3 + 3x^2) + (-4x - 12) \quad \text{grouping}$$

$$= x^2(x + 3) - 4(x + 3)$$

$$= (x + 3)(x^2 - 4) \quad \text{find GCF}$$

$$= (x + 3)(x + 2)(x - 2) \quad \text{factorization}$$

The GCF of the first group $(x^3 + 3x^2)$ is x^2.
Factor it out of the expression to get $x^2(x + 3)$.
The GCF of the second group $(-4x - 12)$ is -4.
Factor it out of the expression to get $-4(x + 3)$.

Consider the same coefficient $\boldsymbol{a}$ in the form of $x^2 + \boldsymbol{a}x + b$.

$$(x + 1)(x + 2)(x + 3)(x + 4) - 8$$

$$= \underline{(x + 1)(x + 4)}\,\underline{(x + 2)(x + 3)} - 8 \quad \text{; grouping}$$

$$= (x^2 + 5x + 4)(x^2 + 5x + 6) - 8$$

$$= (A + 4)(A + 6) - 8 \quad \text{; Substitute } x^2 + 5x \text{ as } A$$

$$= A^2 + 10A + 24 - 8$$

$$= A^2 + 10A + 16$$

$$= (A + 2)(A + 8)$$

$$= (x^2 + 5x + 2)(x^2 + 5x + 8) \quad \text{; replace A with } x^2 + 5x$$

4. If polynomials contain more than 2 variables,

⇒ Rewrite the polynomials in the descending order of powers for the variable with the lowest power. Consider the other terms (which don't contain the variable) as constants. Then factor the pretended constant term.

x has a power of 2.

y has a power of 1.

Because y has the lowest power, choose it for the descending order of powers.

Example

$$x^2 + 2x + xy - y - 3 = (xy - y) + \underline{(x^2 + 2x - 3)} \quad \text{Descending order of powers of } y$$

$$\text{pretended constant term}$$

$$= y(x - 1) + (x^2 + 2x - 3) \quad \text{Factor a group}$$

$$= y(x - 1) + (x + 3)(x - 1) \quad \text{Factor a group}$$

$$= (x - 1)(y + x + 3) \quad \text{Common factor}$$

$$= (x - 1)(x + y + 3)$$

Exercises

#1. Find all factors for the following expressions .

(1) $ab + a + b + 1$

(2) $abx + aby$

(3) $2x^2 + xy$

#2. Factor the following polynomials

(1) $a^2 - ab + a$

(2) $a^2b - ab^2$

(3) $4a - 10$

(4) $2a^3 + 3a^2 - 5a$

(5) $9a^4b^2 - 3a^3b^2 + 12a^2b^2$

(6) $2a(x + y) - 4b(x + y)$

(7) $-12(x - 2y) - 18a(x - 2y)$

(8) $4a(a - b) + 4b(b - a)$

(9) $a^n + a^{n+2}$

(10) $3x^{2n} + 12x^{3n} + 9x^n$

#3. Factor each polynomial using factorization formulas.

(1) $x^2 - 2x + 1$

(2) $9x^2 + 6x + 1$

(3) $4x^2 - 4x + 1$

(4) $x^2 + 10x + 25$

(5) $x^2 + x + \frac{1}{4}$

(6) $x^6 + 6x^3 + 9$

(7) $x^{2n} - 2x^n y^n + y^{2n}$

(8) $x^2 - 1$

(9) $x^2 - 4y^2$

(10) $9x^2 - \frac{1}{4}y^2$

(11) $(x + a)^2 - 36$

(12) $1 - (x + y)^2$

(13) $x^4 - 1$

(14) $x^8 - y^8$

(15) $-16x^2 + 9y^2$

(16) $\frac{1}{9}x^2 - \frac{9}{16}y^2$

(17) $4x^3 y - xy^3$

(18) $x^2 + 4x + 3$

(19) $x^2 - 4x - 5$

(20) $x^2 - 3x + 2$

(21) $x^2 - 2x - 8$

(22) $(x - y)^2 - (x + y)^2$

(23) $(2x - 3)^2 - (x + 1)^2$

(24) $x^2 - \frac{5}{6}x + \frac{1}{6}$

(25) $3x^2 - 4x - 4$

(26) $2x^2 + 3x - 2$

(27) $4x^2 - 2x - 12$

(28) $4x^2 - 10x - 6$

(29) $2x^2 - 3xy - 9y^2$

(30) $x^3y - 16xy^3$

(31) $\frac{1}{3}x^2 - 2 + \frac{3}{x^2}$

(32) $4x^2 - 12xy + 9y^2$

(33) $\frac{1}{3}x^2 - \frac{1}{3}x - 2$

(34) $a^4(x - y) + b^4(y - x)$

(35) $-3a^2 + 3a + 6$

(36) $3ax^2 - 5ax - 2a$

(37) $(a - 1)^2 - 10(a - 1) + 25$

(38) $a^8 - 1$

(39) $(x - 1)(x + 2) - 4$

(40) $4x^2 - 2x - 2 - (x - 1)^2$

#4. Find the value of k that will make the following polynomials perfect square forms.

(1) $x^2 + 5x + k$

(2) $9x^2 - 12x + k$

(3) $2x^2 - 6x + k$

(4) $25x^2 + 4x + k$

(5) $\frac{1}{25}x^2 + kx + 4$

(6) $4x^2 - kx + 25$

(7) $2x^2 + kx + 8$

(8) $9x^2 + (2k - 4)x + 4$

(9) $4x^2 + (k + 5)x + 9y^2$

(10) $k - \frac{1}{4}xy + \frac{1}{4}y^2$

(11) $9x^2 + (k - 1)xy + 25y^2$

(12) $kx^2 + 3x + 9$

(13) $(x + 3)(x - 4) - k$

(14) $(2x + 1)(2x - 4) + k$

(15) $(x - 1)(x - 2)(x + 4)(x + 5) + k$

#5. Find the value of a for the following polynomials. Each polynomial has a given factor.

(1) $x^2 + 2x + a$ has the factor $(x + 3)$.

(2) $3x^2 + ax - 8$ has the factor $(x - 2)$.

(3) $4x^2 + ax - 6$ has the factor $(3 - 2x)$.

(4) $2x^2 + (3a - 1)x - 15$ has the factor $(2x + 3)$.

(5) $2ax^2 - 5x + 2$ has the factor $(3x - 2)$.

#6. Find the value of $a + b$ for any constants a and b.

(1) $3ax^2 - 6x + ab$ has the factor $(3x - 1)^2$.

(2) $ax^2 + 8x + 4b$ has two factors $(3x + 2)$ and $(x - 2)$.

(3) $(4x - 3)^2 - (3x - 2)^2$ has two factors $(ax + 5)$ and $(b - x)$.

(4) $2x^2 + ax - 4$ and $bx^2 - x - 2$ have the same factor $(2x + 1)$.

#7. For any integers $a (\neq 0)$ and b, the length and width of a rectangle are forms of $ax + b$.
Find the perimeter of a rectangle whose area is $6x^2 + 17x + 12$.

#8. Find a possible expression for the width of a rectangle whose area is $12x^2 + 5x - a$, where $a > 0$ and the width is greater than the length.

#9. The area and base of a triangle are $5x^2 + 12x + 4$ and $2x + 4$, respectively. Find the height of the triangle.

#10. The figures $A, B,$ and C (equilateral) have the same area. Find the perimeters of A and C.

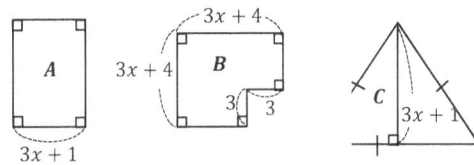

#11. A and B are squares with the lengths a and b, respectively. The perimeter of A is 40 more than the perimeter of B. The difference between their areas is 200. Find the sum of their areas.

#12. Factor each polynomial using any method.

(1) $3a^2b - a^3b - 2ab$

(2) $3a - 3a^2 + 6$

(3) $8a^2b + 6ab - 20b$

(4) $a^5 - 16a$

(5) $a^2(x - y) + b^2(y - x)$

(6) $ab(a - b) + 2a(b - a)^2$

(7) $(a - b)^2 - 2(b - a)^3$

(8) $a(x - y)^2 + b(x - y)$

(9) $4(x + 2)^2 - 3(x + 2) - 10$

(10) $(x - y)(x - y + 3) - 4$

(11) $(2x - y)^2 + 8y(2x - y) + 16y^2$

(12) $2(x - 1)^2 - 3(x - 1)(y + 1) - 9(y + 1)^2$

(13) $a^4 - 5a^2 - 36$

(14) $a^8 - 2a^4 - 8$

(15) $x^2 - xy + 2x - 2y$

(16) $a^2 - ab - b - 1$

(17) $a^3 - a^2 - a + 1$

(18) $a^4 + 3a - 3a^3 - a^2$

(19) $2ab + 1 - a^2 - b^2$

(20) $(x + 2)(x - 1)^2(x - 4) - 10$

(21) $9x^2 - y^2 - 4y - 4$

(22) $a^2 - b^2 + 4b - 4$

(23) $a^2 - 16b^2 - 6a + 9$

(24) $ax^2 - a - bx^2 + b$

(25) $x^2 - xy + x + 2y - 6$

(26) $-2a^2 - 5a + 2ab - b + 3$

(27) $2x^2 - y^2 + xy - 2x + y$

(28) $4a^2 - b^2 - 4a + 4b - 3$

(29) $4a^2 - 4ab + b^2 - c^2$

(30) $a^4 + a^2 + 1$

(31) $a^4 - 6a^2 + 1$

(32) $a^4 - 13a^2 + 4$

(33) $9x^4 + 8x^2 + 4$

#13. Evaluate each expression using factorization.

(1) $99^2 - 1$

(2) $99^2 - 89^2$

(3) $49^2 - 51^2$

(4) $3^8 - 1$

(5) $6^2 - 5^2 + 4^2 - 3^2 + 2^2 - 1$

(6) $\left(1 - \frac{1}{2^2}\right)\left(1 - \frac{1}{3^2}\right)\left(1 - \frac{1}{4^2}\right) \cdots$
$\cdots \left(1 - \frac{1}{99^2}\right)\left(1 - \frac{1}{100^2}\right)$

(7) $3(2^2 + 1)(2^4 + 1)(2^8 + 1) + 1$

(8) $\dfrac{99 \times 101 + 99 \times 2}{101^2 - 4}$

(9) $36 \times 34 - 35 \times 34$

(10) $87 \times 56 + 87 \times 44$

(11) $65^2 - 2 \times 65 \times 35 + 35^2$

(12) $25^2 + 30 \times 25 + 15^2$

(13) $a^2 - 8a - 20$ when $a = 28$

(14) $a^2 + 3a - 54$ when $a = 91$

Algebra

Part II
Expressions

Chapter 7

Quadratic Functions
Level III

CHAPTER
7

Chapter 7 Quadratic Equations

7-1 Quadratic Equations and Their Solutions

1. **Definition**

2. **Solving Quadratic Equations**

   (1) **Using Factorization and The Zero Product Property**

   (2) **Using Perfect Squares**

   (3) **Using Quadratic Formulas**

3. **The Discriminant** : $D = b^2 - 4ac$

4. **The Solution-Coefficient Relationship**

Chapter 7. Quadratic Equations

7-1 Quadratic Equations and Their Solutions

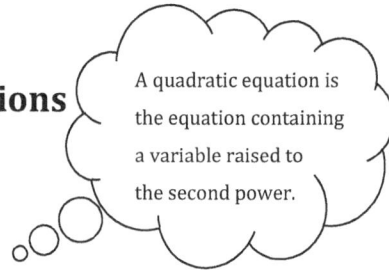

A quadratic equation is the equation containing a variable raised to the second power.

1. Definition

(1) A *quadratic equation* which has transferred all the terms on the right side of the equal sign to the left side of the equal sign is formed by

$$ax^2 + bx + c = 0, \text{ for any constants } a \neq 0, \ b, \text{ and } c.$$

Note : Quadratic equations : $ax^2 + bx + c = 0, \ a \neq 0$

$$ax^2 + bx = 0, \ a \neq 0 \ (\textit{When } c = 0)$$
$$ax^2 + c = 0, \ a \neq 0 \ (\textit{When } b = 0)$$
$$ax^2 = 0, \ a \neq 0 \ (\textit{when } b = 0 \textit{ and } c = 0)$$

(2) The *solution* (*root*) of an unknown variable, x,

must make the quadratic equation ($ax^2 + bx + c = 0, a \neq 0$) always true.

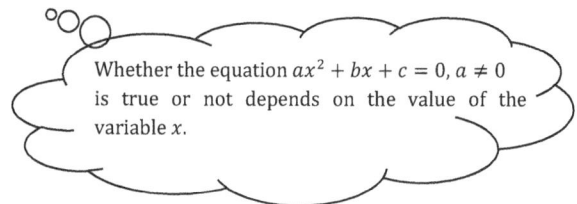

Whether the equation $ax^2 + bx + c = 0, a \neq 0$ is true or not depends on the value of the variable x.

Example

$$x^2 - 2x + 1 = 0$$

If $x = 1$, then $1 - 2 + 1 = 0$ (this makes the quadratic equation true.)

∴ $x = 1$ is a solution.

If $x = -1$, then $1 + 2 + 1 \neq 0$ (this makes the quadratic equation untrue.)

∴ $x = -1$ is not a solution.

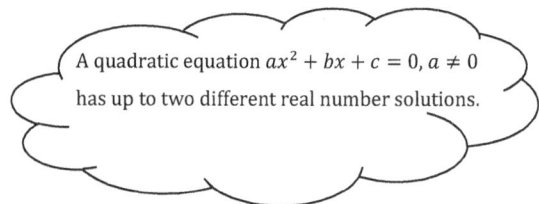

A quadratic equation $ax^2 + bx + c = 0, a \neq 0$ has up to two different real number solutions.

2. Solving Quadratic Equations

> To apply the zero product property,
> if $ax^2 + bx + c = k$,
> $\Rightarrow ax^2 + bx + c - k = 0$
> ($\because$ One side of equal sign must equal 0.)

(1) Using Factorization and The Zero Product Property

1) The Zero product property

> For any expressions or numbers A and B,
>
> $AB = 0 \underset{\text{if and only if}}{\Longleftrightarrow} A = 0 \text{ or } B = 0$

$$Note: \ A = 0 \ or \ B = 0 \Leftrightarrow \begin{cases} A = 0 \ and \ B \neq 0 \\ A \neq 0 \ and \ B = 0 \\ A = 0 \ and \ B = 0 \end{cases}$$

> $AB \neq 0 \Leftrightarrow A \neq 0 \text{ and } B \neq 0$

2) Steps for Solving Quadratic Equations

Step 1 : Simplify the equation as a form of $ax^2 + bx + c = 0, a \neq 0$

Step 2 : Factor the left side of the equation into the form of $a(x - \alpha)(x - \beta) = 0$

Step 3 : Solve the equation using the zero product property ;

$\quad (x - \alpha) = 0 \text{ or } (x - \beta) = 0 \ ; \ x = \alpha \ \text{ or } x = \beta$

Example

> $ax^2 + bx + c = 0 \Leftrightarrow (mx + p)(nx + q) = 0,$
> (where $mn = a, \ mq + np = b, \ pq = c$)
> $\Rightarrow mx + p = 0 \text{ or } nx + q = 0$
> $\Rightarrow mx = -p \text{ or } nx = -q$
> $\Rightarrow x = -\dfrac{p}{m} \text{ or } x = -\dfrac{q}{n}$

$2x^2 - 4x - 6 = 0 \underset{\text{Step 1}}{\Longrightarrow} 2(x^2 - 2x - 3) = 0$

$\underset{\text{Step 2}}{\Longrightarrow} 2(x - 3)(x + 1) = 0$

$\underset{\text{Step 3}}{\Longrightarrow} x = 3 \ \text{ or } \ x = -1$

3) Double roots :

If a quadratic equation has a same factor twice, then the equation has a double root.

$ax^2 + bx + c = 0 \Rightarrow a(x - \underline{\alpha})(x - \underline{\alpha}) = 0$: These are two identical factors.

$$\Rightarrow a(x - \alpha)^2 = 0$$

$$\Rightarrow x - \alpha = 0 \quad (\because a \neq 0)$$

$$\Rightarrow x = \alpha \quad \text{(the double root)}$$

$a(x - \alpha)^2 = 0$

$\Rightarrow a(x - \alpha)(x - \alpha) = 0$

$\Rightarrow (x - \alpha)(x - \alpha) = 0 \quad (\because a \neq 0)$

$\Rightarrow x = \alpha$ or $x = \alpha$ ($\because$ by the zero product property)

$\Rightarrow x = \alpha$: the double root (when you get the same answer twice)

Example 1

$x^2 + 4x + 4 = 0 \Rightarrow (x + 2)^2 = 0 \Rightarrow x = -2$ (the double root)

$a(\quad)^2 = \mathbf{0}$

Example 2

If the equation $x^2 - 3x + k = 0$ has a double root, $k = \left(\dfrac{-3}{2}\right)^2 = \dfrac{9}{4}$.

So, $x^2 - 3x + k = x^2 - 3x + \dfrac{9}{4} = \left(x - \dfrac{3}{2}\right)^2 = 0$

$\therefore x = \dfrac{3}{2}$ (the double root)

$x^2 + ax + \left(\dfrac{a}{2}\right)^2 = \left(x + \dfrac{a}{2}\right)^2 = \mathbf{0}$

constant $= \left(\dfrac{x-\text{coefficient}}{2}\right)^2$, when x^2-coefficient $= 1$

① $a(x - \alpha)^2 = 0$, $a \neq 0 \Rightarrow x = \alpha$ (the double root)

② To find a double root for the quadratic equation $a(x - p)^2 = q$,

$q = 0$ (q must equal 0). So, we get $x = p$ (the double root)

Note : ① *If the quadratic equation $x^2 = q$ has solutions, $q \geq 0$ (q must be positive or zero).*

② *If the quadratic equation $x^2 = q$ has 2 different solutions, $q > 0$ (q must be positive).*

③ *If the quadratic equation $x^2 = q$ has a double root, $q = 0$ (q must be zero).*

> If the solution for the quadratic equation $x^2 = q$ does not exist (no solution), $q < 0$ (q must be negative).

(2) Using Perfect Squares

Solve the equation in the form of a perfect square $(x + p)^2 = q$

1) Square roots

If the quadratic equation $ax^2 + bx + c = 0$ is formed by $(x + p)^2 = q$,

$\Rightarrow$ The solution is $\begin{cases} x = -p \pm \sqrt{q} & , \ q > 0 \\ x = -p \ \text{(double root)}, \ q = 0 \\ \text{no solution} & , \ q < 0 \end{cases}$

Note : (1) $x^2 = p, \ p \geq 0 \ \Rightarrow \ x = \pm\sqrt{p}$

(2) $ax^2 = p, \ a \neq 0, \ \frac{p}{a} \geq 0 \ \Rightarrow \ x^2 = \frac{p}{a} \ \Rightarrow \ x = \pm\sqrt{\frac{p}{a}}$

(3) $(x + p)^2 = q, \ q \geq 0 \ \Rightarrow \ x + p = \pm\sqrt{q} \ \Rightarrow \ x = -p \pm \sqrt{q}$

(4) $(ax + p)^2 = q, \ a \neq 0, \ q \geq 0 \ \Rightarrow \ ax + p = \pm\sqrt{q} \ \Rightarrow \ x = \frac{-p \pm \sqrt{q}}{a}$

> (1) Number $\begin{cases} \text{Real number} \begin{cases} \text{positive} \\ 0 \\ \text{negative} \end{cases} \\ \text{Complex number} \end{cases}$
>
> (Real number)$^2 \geq 0$ but ($\underbrace{\text{imaginary number}}_{\text{not a Real number}}$)$^2 < 0$
>
> (2) i = a unit of imaginary numbers
>
> $i = \sqrt{-1}$; $i^2 = -1 < 0$
>
> (3) If there is no condition about the number,
>
> then we consider only real number solutions not complex number solutions.

2) Steps for Solving Quadratic Equations Using Perfect Squares

Step 1. Make the coefficient of x^2 equal to 1.

$$ax^2 + bx + c = 0 \quad \Rightarrow \quad a\left(x^2 + \frac{b}{a}x + \frac{c}{a}\right) = 0$$

Step 2. Divide by a on both sides or use the zero product property.

$$\Rightarrow \ x^2 + \frac{b}{a}x + \frac{c}{a} = 0$$

Step 3. Transfer the constant to the right side of the equal sign by subtracting $\frac{c}{a}$ from both sides.

$$\Rightarrow \ x^2 + \frac{b}{a}x = -\frac{c}{a}$$

Step 4. Square half of the x-coefficient. Add the result to both sides of the equal sign.

$$x^2 + \frac{b}{a}x + \left(\frac{b}{2a}\right)^2 = -\frac{c}{a} + \left(\frac{b}{2a}\right)^2$$

$$\Rightarrow \left(x + \frac{b}{2a}\right)^2 = -\frac{c}{a} + \left(\frac{b}{2a}\right)^2 \quad \text{form of perfect square, } (x+p)^2 = q$$

OR

$$x^2 + \frac{b}{a}x = \left(x + \frac{1}{2}\cdot\frac{b}{a}\right)^2 - \left(\frac{b}{2a}\right)^2 = -\frac{c}{a}$$

$$\Rightarrow \left(x + \frac{1}{2}\cdot\frac{b}{a}\right)^2 = -\frac{c}{a} + \left(\frac{b}{2a}\right)^2 \quad \text{form of perfect square, } (x+p)^2 = q$$

Step 5. Solve the equation using square root.

$$\frac{1}{2}\cdot\frac{b}{a} = \frac{1}{2}\cdot \text{(the coefficient of } x\text{)}$$

$$x^2 + mx = \left(x + \frac{1}{2}m\right)^2 - \left(\frac{1}{2}m\right)^2$$

$$; \text{Since } (x \pm m)^2 = x^2 \pm 2mx + m^2,$$

$$\left(x + \frac{1}{2}m\right)^2 = x^2 + 2\cdot\frac{1}{2}mx + \left(\frac{1}{2}m\right)^2 = x^2 + mx + \left(\frac{1}{2}m\right)^2$$

$$\therefore \ x^2 + mx = \left(x + \frac{1}{2}m\right)^2 - \left(\frac{1}{2}m\right)^2$$

Example

$2x^2 + 5x + 3 = 0$

$\xRightarrow{\qquad\qquad}$ $2\left(x^2 + \dfrac{5}{2}x + \dfrac{3}{2}\right) = 0$

$\underset{\text{divide by 2}}{\xRightarrow{\qquad\qquad}}$ $x^2 + \dfrac{5}{2}x + \dfrac{3}{2} = 0$

$\underset{\text{transfer constant}}{\xRightarrow{\qquad\qquad}}$ $x^2 + \dfrac{5}{2}x = -\dfrac{3}{2}$: move the constant to the right side of the equation

$\xRightarrow{\qquad\qquad}$ $\left(x + \dfrac{1}{2}\cdot\dfrac{5}{2}\right)^2 - \left(\dfrac{1}{2}\cdot\dfrac{5}{2}\right)^2 = -\dfrac{3}{2}$: to find the perfect square

$\xRightarrow{\qquad\qquad}$ $\left(x + \dfrac{5}{4}\right)^2 = -\dfrac{3}{2} + \left(\dfrac{5}{4}\right)^2$

$\xRightarrow{\qquad\qquad}$ $\left(x + \dfrac{5}{4}\right)^2 = \dfrac{1}{16}$: perfect square form

$\underset{\text{use the square root}}{\xRightarrow{\qquad\qquad}}$ $x + \dfrac{5}{4} = \pm\sqrt{\dfrac{1}{16}}$: to solve the equation for x

$\xRightarrow{\qquad\qquad}$ $x = -\dfrac{5}{4} \pm \dfrac{1}{4}$

$\xRightarrow{\qquad\qquad}$ $x = -1$ or $-\dfrac{6}{4}\left(= -\dfrac{3}{2}\right)$: the solution to the equation

(3) Using Quadratic Formulas

If an equation is not factorized, solve it using quadratic formulas.

1) Quadratic Formula I

$$ax^2 + bx + c = 0, \quad a \neq 0$$
$$\Rightarrow \quad x = \frac{-b \pm \sqrt{b^2 - 4ac}}{2a}, \quad b^2 - 4ac \geq 0$$

2) Quadratic Formula II (When $b = 2b'$ is an even number)

$$ax^2 + 2b'x + c = 0, \quad a \neq 0$$
$$\Rightarrow \quad x = \frac{-b' \pm \sqrt{b'^2 - ac}}{a}, \quad b'^2 - ac \geq 0$$

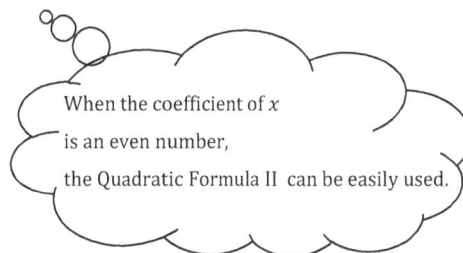

When the coefficient of x is an even number, the Quadratic Formula II can be easily used.

Note : $ax^2 + bx + c = 0, \ a \neq 0$

$$\Rightarrow \ a\left(x^2 + \frac{b}{a}x + \frac{c}{a}\right) = 0 \ \Rightarrow \ x^2 + \frac{b}{a}x + \frac{c}{a} = 0 \ \Rightarrow \ x^2 + \frac{b}{a}x = -\frac{c}{a}$$

$$\Rightarrow \ \left(x + \frac{1}{2}\cdot\frac{b}{a}\right)^2 - \left(\frac{b}{2a}\right)^2 = -\frac{c}{a} \ \Rightarrow \ \left(x + \frac{b}{2a}\right)^2 = \frac{b^2 - 4ac}{4a^2}$$

$$\Rightarrow \ x + \frac{b}{2a} = \pm\sqrt{\frac{b^2 - 4ac}{4a^2}} \ \Rightarrow \ x = -\frac{b}{2a} \pm \sqrt{\frac{b^2 - 4ac}{4a^2}}$$

$$\Rightarrow \ x = \frac{-b \pm \sqrt{b^2 - 4ac}}{2a}$$

Example 1

$x^2 + 5x + 3 = 0$ (This equation can't be factorized.)

$$\xrightarrow[a=1, b=5, c=3]{} \quad x = \frac{-5 \pm \sqrt{(5)^2 - 4\cdot 1 \cdot 3}}{2\cdot 1} = \frac{-5 \pm \sqrt{13}}{2} \quad \text{Using Quadratic Formula I}$$

$$\therefore \ x = \frac{-5 + \sqrt{13}}{2} \ \text{ or } \ x = \frac{-5 - \sqrt{13}}{2}$$

Example 2

$4x^2 + 4x - 3 = 0$

① Using Quadratic Formula I

$$\xrightarrow[a=4, b=4, c=-3]{} \quad x = \frac{-4 \pm \sqrt{(4)^2 - 4\cdot 4 \cdot(-3)}}{2\cdot 4} = \frac{-4 \pm \sqrt{64}}{8} = \frac{-4 \pm 8}{8}$$

$$\therefore \ x = \frac{1}{2} \ \text{ or } \ x = -\frac{3}{2}$$

② Using Quadratic Formula II

$$\xrightarrow[a=4, b\prime=2, c=-3]{} \quad x = \frac{-2 \pm \sqrt{(2)^2 - 4\cdot(-3)}}{4} = \frac{-2 \pm \sqrt{16}}{4} = \frac{-2 \pm 4}{4}$$

$$\therefore \ x = \frac{1}{2} \ \text{ or } \ x = -\frac{3}{2}$$

③ Using Factorization

$$4x^2 + 4x - 3 = (2x - 1)(2x + 3) = 0$$

$$\therefore \ x = \frac{1}{2} \ \text{ or } \ x = -\frac{3}{2}$$

④ Using a Perfect Square

$$4x^2 + 4x - 3 = 0 \ \Rightarrow \ 4\left(x^2 + x - \frac{3}{4}\right) = 0 \ \Rightarrow \ x^2 + x - \frac{3}{4} = 0$$

$$\Rightarrow \ \left(x + \frac{1}{2}\right)^2 - \frac{1}{4} = \frac{3}{4} \ \Rightarrow \ \left(x + \frac{1}{2}\right)^2 = 1$$

$$\Rightarrow \ x + \frac{1}{2} = \pm 1 \ \Rightarrow \ x = -\frac{1}{2} \pm 1$$

$$\therefore \ x = \frac{1}{2} \ \text{ or } \ x = -\frac{3}{2}$$

The solutions of a quadratic equation depend on the sign of the number D under the symbol $\sqrt{}$.

In order to have possible solutions, $D \geq 0$.

3. The Discriminant $D = b^2 - 4ac$

For a quadratic equation of the form $ax^2 + bx + c = 0$, the solution is

$$x = \frac{-b \pm \sqrt{b^2 - 4ac}}{2a} \ , \ \ b^2 - 4ac \geq 0$$

$D = b^2 - 4ac$ is called the *Discriminant* of the quadratic equation and D determines the number of solutions for the equation.

(1) $D > 0 \Rightarrow$ 2 different solutions $\quad x = \dfrac{-b + \sqrt{b^2 - 4ac}}{2a} \quad$ or $\quad x = \dfrac{-b - \sqrt{b^2 - 4ac}}{2a}$

(2) $D = 0 \Rightarrow$ the same solution twice; only 1 solution (double root) $\quad x = \dfrac{-b}{2a}$

(3) $D < 0 \Rightarrow$ no real number solution exist

A quadratic equation $ax^2 + bx + c = 0$ has a double root.

$\Leftrightarrow a(x + p)^2 = \mathbf{0}, a \neq 0$

$\Leftrightarrow D = b^2 - 4ac = 0$

$a + b\sqrt{k} = 0$

$\Rightarrow a = 0$ and $b\sqrt{k} = 0$

$\Rightarrow a = 0$ and $b = 0 \ (\because \sqrt{k} \neq 0)$

4. The Solution-Coefficient Relationship

The solutions and the coefficients of a quadratic equation are related to the obtaining the sum and the product of the solutions. We can also obtain a quadratic equation from the solutions.

(1) Given a quadratic equation $ax^2 + bx + c = 0$, suppose the solutions are $x = \alpha$ and $x = \beta$. Then we obtain :

> If $\alpha\beta = \frac{c}{a} < 0 \ \Rightarrow \ ac < 0$
> $\Rightarrow \ D = b^2 - 4ac > 0 \ (\because b^2 > 0 \text{ and } -(ac) > 0 \ )$
> $\Rightarrow \ $ 2 different solutions exist.

$$1) \ \alpha + \beta = -\frac{b}{a} \quad \text{(the sum of the solutions)}$$

$$2) \ \alpha \cdot \beta = \frac{c}{a} \quad \text{(the product of the solutions)}$$

Note : $ax^2 + bx + c = 0, \ a \neq 0$

$$\Rightarrow \ x = \frac{-b+\sqrt{b^2-4ac}}{2a} \ or \ x = \frac{-b-\sqrt{b^2-4ac}}{2a}$$

Let $\alpha = \dfrac{-b+\sqrt{b^2-4ac}}{2a}$ and $\beta = \dfrac{-b-\sqrt{b^2-4ac}}{2a}$

Then, $\alpha + \beta = \dfrac{-b+\sqrt{b^2-4ac}}{2a} + \dfrac{-b-\sqrt{b^2-4ac}}{2a} = \dfrac{-2b}{2a} = -\dfrac{b}{a}$

$$\alpha \cdot \beta = \frac{-b+\sqrt{b^2-4ac}}{2a} \cdot \frac{-b-\sqrt{b^2-4ac}}{2a} = \frac{(-b)^2-\left(\sqrt{b^2-4ac}\right)^2}{4a^2} = \frac{b^2-(b^2-4ac)}{4a^2} = \frac{4ac}{4a^2} = \frac{c}{a}$$

> $(\alpha \pm \beta)^2 = \alpha^2 \pm 2\alpha\beta + \beta^2$
> $\alpha^2 + \beta^2 = (\alpha + \beta)^2 - 2\alpha\beta$
> $(\alpha + \beta)^2 = (\alpha - \beta)^2 + 4\alpha\beta$
> $(\alpha - \beta)^2 = (\alpha + \beta)^2 - 4\alpha\beta$
> $\alpha^2 - \beta^2 = (\alpha + \beta)(\alpha - \beta)$

(2) If the solutions are given, then a quadratic equation can be obtained.

1) $\begin{pmatrix} x^2\text{-coefficient} = a \ (\neq 0) \\ \text{Solutions} : x = \alpha \text{ and } x = \beta \end{pmatrix}$ $\Rightarrow$ The quadratic equation is $a(x - \alpha)(x - \beta) = 0$

2) $\begin{pmatrix} x^2\text{-coefficient} = a \ (\neq 0) \\ \text{Solution} : x = \alpha \text{ (the double root)} \end{pmatrix}$ $\Rightarrow$ The quadratic equation is $a(x - \alpha)^2 = 0$

3) $\begin{pmatrix} x^2\text{-coefficient} = a \ (\neq 0) \\ \text{The sum of the solutions} = p, \\ \text{The product of the solutions} = q \end{pmatrix}$ $\Rightarrow$ The quadratic equation is $a(x^2 - px + q) = 0$

> $a(x - \alpha)(x - \beta) = 0 \Leftrightarrow a(x^2 - (\alpha + \beta)x + \alpha\beta) = 0$
> where $(\alpha + \beta)$ is the sum of the solutions and
> $\alpha\beta$ is the product of the solutions

(3) If a quadratic equation has a solution $\alpha + \beta\sqrt{m}$, then the other solution is $\alpha - \beta\sqrt{m}$.

Note : $ax^2 + bx + c = 0, a \neq 0$

$\Rightarrow \quad x = \dfrac{-b+\sqrt{b^2-4ac}}{2a} \quad (\textit{form of } \alpha + \beta\sqrt{m})$

$or \quad x = \dfrac{-b-\sqrt{b^2-4ac}}{2a} \quad (\textit{form of } \alpha - \beta\sqrt{m})$

Example 1

With given solutions, we can find a quadratic equation.

If the x^2-coefficient is 3 and the solutions are $x = 1$ and $x = 2$, then the quadratic equation is

$3(x-1)(x-2) = 0 \ ; \ 3(x^2 - 3x + 2) = 0 \ $ or

$3(x^2 - (1+2)x + (1 \cdot 2)) = 0 \ ; \ 3(x^2 - 3x + 2) = 0$

Example 2

With the coefficients of a quadratic equation, we can find the solutions.

If the quadratic equation $x^2 - 7x + 12 = 0$ has solutions $x = \alpha$ and $x = \beta$, then

$\alpha + \beta = -\dfrac{-7}{1} = 7 \ $ and $\ \alpha \cdot \beta = \dfrac{12}{1} = 12$

$\therefore \ (\alpha - \beta)^2 = (\alpha + \beta)^2 - 4\alpha\beta = 7^2 - 4 \cdot 12 = 1 \ ; \ \alpha - \beta = \pm 1$

$\therefore \quad \alpha + \beta = 7$

$+) \ \underline{\alpha - \beta = \pm 1}$

$\quad 2\alpha \quad = 7 \pm 1 \ ; \ \alpha = 4 \ $ or $\ \alpha = 3$

Therefore, the solutions are $x = 4$ and $x = 3$.

Exercises

#1. State whether each expression is a quadratic equation (Yes) or is not a quadratic equation(No).

   (1) $x^2 = 5$

   (2) $x(x + 3) = 0$

   (3) $x^2 = (x + 3)^2$

   (4) $x^3 + x^2 + 1 = 0$

   (5) $2x^3 + x^2 = 3x + 2x^3 + 5$

   (6) $\frac{1}{x^2} + \frac{1}{x} + 3 = 0$

   (7) $(x + 1)^2 - (x - 1)^2 + 3 = 0$

   (8) $2x^2 + 5x = (x + 1)(x + 2)$

   (9) $x^2 + 4x + 4 = 2(x + 2)^2 - 2$

   (10) $x^2 - 5x + 6$

(11) $3x^2 + 5x + 1 = 2x^2 - 6x + 4$

(12) $\frac{x^2}{2} + 4x = 4x + 4$

(13) $\frac{2}{x^2} + 2x = 3$

(14) $x^2 + 1 = (x + 1)^2$

(15) $x^2 + 2x + 1 = 2(x + 1)^2$

(16) $x^2 = 0$

(17) $x(2x + 1) = 3x(x + 2)$

(18) $(x + 4)^2$

(19) $x^2(x - 1) = x(x^2 + x - 1)$

(20) $x^2 + 3x = x^2 + 3$

#2. The following equations are quadratic equations. Find the condition for constants a and b.

(1) $(x + 1)(ax + 2) = 2x^2 + 5$

(2) $2(x^2 + 2x + 1) = (x + 2)(5 - ax)$

(3) $(3x - 1)(ax + 2) = 5 - bx^2$

(4) $(2x + 1)(3x + 2) = (ax + 2)(bx - 3)$

(5) $(2a + b)x^2 + ax + b = 0$

(6) $a^2x^2 + bx + 5 = 5$

#3. Find the value of $a + b + c$ for the quadratic equation $ax^2 + bx + c = 0$, in which a is the smallest positive number.

(1) $2(x - 1)^2 = (x + 1)^2 + 5$

(2) $(3x + 1)(x - 2) = 2 - x^2$

(3) $3(x + 1)^2 = 3(x + 1)$

(4) $x^2 = x$

(5) $2x(x - 1) = x^2 - 2$

#4. Find the sum of the solutions for each quadratic equation using factorization.

(1) $x^2 - 2x - 3 = 0$

(2) $2x^2 - 7x + 5 = 0$

(3) $-3x^2 + 6x = 0$

(4) $2x^2 + 2x - 4 = 0$

(5) $x(x + 5) = 6$

(6) $x^2 = \frac{x+1}{2}$

#5. Each of the following quadratic equations has a solution $x = \alpha$. For each equation, find the value of constant a and the other solution $x = \beta$ for the equation.

(1) $x^2 + ax - 4 = 0, \ x = 1$

(2) $3x^2 - ax + a = 0, \ x = -1$

(3) $2x^2 - x + a = 0, \ x = -1$

(4) $ax^2 - 2x - 3 = 0, \ x = -2$

(5) $ax^2 - (a-1)x - 6 = 0, \ x = 2$

#6. Each of the following quadratic equations has the solution $x = \alpha$. Find the value of the given expression for each equation.

(1) $\alpha - \frac{1}{\alpha}$ for $x^2 - 2x + 1 = 0$

(2) $\alpha^2 + \frac{1}{\alpha^2}$ for $2x^2 + 3x - 2 = 0$

(3) $(\alpha + \frac{1}{\alpha})^2$ for $x^2 - 3x - 1 = 0$

(4) $\alpha^2 + \frac{9}{\alpha^2}$ for $3x^2 + 2x - 9 = 0$

(5) $\frac{\alpha-1}{\alpha+1} - \frac{\alpha+1}{\alpha-1}$ for $2x^2 - 6x - 2 = 0$

(6) $\alpha^2 + \alpha - \frac{1}{\alpha} + \frac{1}{\alpha^2}$ for $x^2 - 4x - 1 = 0$

#7. Find the value of the given expression for the following quadratic equations. Each equation has two solutions (α, β) where $\alpha > \beta$.

(1) $\alpha + \beta$ for $x^2 - 5x - 6 = 0$

(2) $\alpha - \beta$ for $x^2 + 3x - 10 = 0$

(3) $\dfrac{\alpha+\beta}{\alpha-\beta}$ for $12x^2 + 5x - 3 = 0$

(4) $\alpha^2 - \beta^2$ for $6x^2 - x - 1 = 0$

#8. Solve the following quadratic equations using square roots

(1) $2x^2 = 8$

(2) $9x^2 - 5 = 0$

(3) $3(x - 1)^2 = 15$

(4) $(2x + 5)^2 - 3 = 0$

(5) $4(x - 2)^2 - 1 = 0$

#9. Solve the following quadratic equations using perfect squares

(1) $x^2 - 3x - 3 = 0$

(2) $2x^2 + 5x = 7$

(3) $-x^2 - 3x + 5 = 0$

(4) $3x^2 - 4x + 1 = 0$

#10. Find the constant k for the following quadratic equations with a double root

(1) $(3x - 4)^2 - k^2 = 0$

(2) $x^2 - kx + 5 = 0$

(3) $x^2 + 2x + k^2 = 0$

(4) $kx^2 + 3x + 2 = 0$

(5) $2x^2 + 3x + k - 5 = 0$

(6) $x^2 + kx + (k - 1) = 0$

(7) $\frac{1}{3}x^2 + (k + 1)x + 8 = 0$

#11. Find the value of $p + q$ for the following quadratic equations with the solution $x = p \pm \sqrt{q}$

(1) $-2x^2 + 5x + 1 = 0$

(2) $3(x - 1)^2 = 4$

(3) $-(x + 1)^2 + 5 = 0$

#12. Find the constant a or the range of a for the following quadratic equations with a condition

(1) $(x + 1)^2 = a + 2$ has no solution.

(2) $x^2 + 3x + 3a = 0$

has two different solutions.

(3) $ax^2 + x + 2 = 0$ has one solution.

(4) $3x^2 - x + a = 0$ has no solution.

(5) $x^2 + (a + 1)x + \frac{a+3}{2} = 0$ has a double root.

#13. Find the value of the given expression for the following quadratic equations with two solutions (α, β)

(1) $\alpha\beta$ for $2(x + 3)^2 - 3 = 0$

(2) $\alpha + \beta$ for $-(x + 4)^2 + 5 = 0$

(3) $\alpha^2 + \beta^2$ for $3x^2 - x - 1 = 0$

#14. Find two constants (a and b) for the following quadratic equations with a condition

(1) $x^2 + x + a = 0$ has two different solutions, $x = 2$ and $x = b$.

(2) $x^2 + 2ax + b = 0$ has a double root $x = 3$.

(3) $x^2 + ax + b = 0$ has a solution $x = 1 + \sqrt{2}$.

(4) $x^2 - ax - 2b^2 = 0$ has two different solutions, $x = 4 \pm \sqrt{2a}$.

#15. Solve the following quadratic equations using quadratic formulas

(1) $x^2 - 2x - 4 = 0$

(2) $3x^2 + 5x - 1 = 0$

(3) $5x^2 - 2x - 1 = 0$

(4) $-2x^2 + 3x + 5 = 0$

(5) $\frac{1}{2}x^2 - 3x + 2 = 0$

(6) $\frac{1}{6}x^2 - 0.5x + \frac{1}{4} = 0$

(7) $(x + 1)^2 = 3(x + 2)$

(8) $(x + 2)^2 + 3(x + 2) - 2 = 0$

(9) $-\frac{(x-1)^2}{2} + x = 0.4(x + 1)$

(10) $(x + 3)(2x + 6) = 5$

#16. Find the value of the given expression for the following quadratic equations with a solution.

(1) $a + b$

for $(2x + 1)^2 = 3$ with $x = a \pm b\sqrt{3}$

(2) $a - b$

for $2x^2 - 8x + 1 = 0$ with $x = \dfrac{a \pm 3\sqrt{b}}{6}$

(3) ab

for $ax^2 + 5x + 2 = 0$ with $x = \dfrac{-5 \pm 2\sqrt{b}}{4}$

(4) $\dfrac{b}{a}$

for $3x^2 - 5x + 1 = 0$ with $x = a \pm \sqrt{b}$

(5) $\dfrac{a+b}{ab}$

for $x^2 - 3x + 1 = 0$ with $x = \dfrac{a \pm 2\sqrt{b}}{2}$

(6) $\dfrac{a-b}{a^2-b^2}$

for $ax^2 + 3x - 3b = 0$ with $x = -1 \pm \sqrt{5}$

#17. Identify the number of solutions for each quadratic equation.

(1) $x^2 + 2x - 3 = 0$

(2) $-x^2 + x - 5 = 0$

(3) $4x^2 - 4x + 1 = 0$

(4) $kx^2 - (k+5)x + 1 = 0$

(5) $3x^2 - x - k^2 = 0$

(6) $x^2 - 4kx + 5k^2 + 1 = 0$

#10. Find the value of a or range of a for the following quadratic equations with a condition (Use the discriminant D)

(1) $x^2 + 5x + a = x + 2$ has no solution.

(2) $(a+3)x^2 - 2ax + a - 1 = 0$

has two different solutions.

(3) $x^2 + 3ax - 2a + 3 = 0$

has only one solution.

(4) $x^2 + ax + a + 2 = 0$ has a double root and $x^2 + 4ax + (2a-1)^2 = 0$ has two different solutions.

(5) $2x^2 + (2a-1)x + a^2 + \frac{1}{4} = 0$

has solutions.

(6) $(a-1)x^2 + 2(a-1)x + (a+1) = 0$ has solutions.

#19. A quadratic equation $3x^2 + 5x - 2 = 0$ has two solutions, $x = \alpha$ and $x = \beta$. Find the value of the given expressions.

(1) $\alpha + \beta$

(2) $\alpha^2 + \beta^2$

(3) $\alpha - \beta$

(4) $\alpha^2 - \beta^2$

(5) $\frac{1}{\alpha} + \frac{1}{\beta}$

#20. A quadratic equation $x^2 + 3kx + 2k^2 - 4k - 1 = 0$ has two solutions, $x = \alpha$ and $x = \beta$. Find the value of the given expressions in terms of k for (1) through (5). Find the value of k for (6).

(1) $\alpha + \beta$

(2) $\alpha\beta$

(3) $\alpha^2 + \beta^2$

(4) $(\alpha - \beta)^2$

(5) $\frac{\beta}{\alpha} + \frac{\alpha}{\beta}$

(6) k if $\frac{1}{\alpha} + \frac{1}{\beta} = 1$

#21. Find the solution for the quadratic equation

$$ax^2 + (b-1)x + 4 = 0$$

(1) When the quadratic equation

$2x^2 + (a-1)x + b = 0$ has two solutions, $\frac{1}{2}$ and $\frac{1}{3}$.

(2) When the quadratic equation

$3ax^2 + 8bx + 3 = 0$ has a double root -2.

(3) When the quadratic equation

$ax^2 + 3ax - 4 = 0$ has two solutions, b and $b+1$.

(4) When the quadratic equation

$ax^2 + 3x + b = 0$ has two solutions, α and β, which satisfy the conditions $\alpha + \beta = -2$ and $\alpha\beta = 4$.

(5) When the quadratic equation $x^2 + ax + 3 = 0$ has two different solutions. The one of the solutions is $x = -2 + 3\sqrt{b}$.

#22. $x^2 + ax + b = 0$ has two solutions, -2 and -3. Find the value of $\alpha^2 + \beta^2$ for $x^2 - bx - a = 0$ which has solutions, α, β.

#23. Create a 1 x^2-coefficient quadratic equation which has two solutions, $\alpha + \beta$ and $\alpha\beta$, where α and β are both solutions of $x^2 + 2x - 3 = 0$.

#24. Find the range of a for the following quadratic equations with a condition

(1) The quadratic equation

$x^2 - 3x + 2a = 0$ has two different positive solutions.

(2) The quadratic equation

$ax^2 + 2x + 3 = 0$ has two different negative solutions.

(3) The quadratic equation

$x^2 - 4x + 3a = 0$ has two different solutions (α and β) with opposite signs.

#25. The following quadratic equations have only one solution. Find the solution (a double root) for each.

(1) $x^2 + kx + 2k - 3 = 0$

(2) $(k + 2)x^2 - 2kx + k + 1 = 0$

(3) $x^2 + (k + 2)x + k^2 - k + 2 = 0$

#26. An n-sided polygon has $\frac{n(n-3)}{2}$ diagonals. Find a polygon that has 20 diagonals.

#27. For three consecutive positive integers, the square of the biggest number is 12 less than the sum of the squares of the other numbers. Identify the biggest number.

#28. The product of two consecutive odd numbers is 99. Find the sum of the numbers.

#29. The sum of two positive numbers is 34 and their product is 225. Identify the two numbers.

#30. Nichole wants to produce a x^2 % of salt solution after mixing 40 ounces of a 10% of salt solution with 40 ounces of a x% salt solution. Find the value of x.

#31. The difference between two positive integers is 2 and their product is 255. Find the sum of the numbers.

#32. The area of a square A is 121 square inches. Each side of square A is 3 inches longer than that of square B. Find the perimeter of the Square B.

#33. The perimeter and the area of a rectangle are 26 inches and 40 square inches, respectively. Find the difference between the length and width of the rectangle's sides (in this case, length will be longer than width).

#34. Richard throws a ball upward with a beginning speed v of 60 feet per second. The formula for the height in feet after t seconds is $h = vt - 5t^2$

    (1) At what time will the height 100 feet?

    (2) When will the ball reach the ground again?

    (3) What height will the ball reach in 4 seconds?

    (4) What is the maximum height the ball will reach?

Algebra

**Part II
Expressions**

Chapter 8

**Rational Expressions
Level III**

CHAPTER 8

Chapter 8   Rational Expressions (Algebraic Functions)

Chapter 8. Rational Expressions (Algebraic Functions)

8-1 Simplifying Rational Expressions

1. Rational Numbers and Rational Expressions

(1) *A rational number* is a fraction whose numerator and denominator are both integers and denoted by

$$\frac{a}{b} \quad \text{for any integers } a \text{ and } b(\neq 0).$$

(2) *A polynomial* is the sum of a number of terms, each of which is the product of numbers and letters. *Rational expression* is a fraction whose numerator and denominator are both polynomials and denoted by

$$\frac{P(x)}{Q(x)} \quad \text{for any polynomials } P(x) \text{ and } Q(x)(\neq 0).$$

Note 1.   *Rational functions:*

$y = \frac{1}{x}$ *is not defined at* $x = 0$.

$y = \frac{x+1}{x+2}$ *is not defined at* $x = -2$.

$y = \frac{x^2}{(x-1)(x+3)}$ *is not defined at either* $x = 1$ *or at* $x = -3$.

Note 2.

$y = \frac{x}{x}$ *is not defined at* $x = 0$, *but for other variables of* x, $\frac{x}{x} = 1$.

Thus, the two expressions $\frac{x}{x}$ *and* 1 *are consequently not identical.*

Similarly, $y = \frac{x(x+1)}{x+1}$ *and* $y = x$ *are not the same rational functions.*

($\because$ Both functions have the same variables for $x \neq -1$.

    But $y = \frac{x(x+1)}{x+1}$ *is not defined at* $x = -1$, *whereas* $y = x$ *has the value* -1.)

2. Reducing Fractions to Lowest Terms

To reduce fractions to lowest terms, simplify the rational expressions and identify restrictions on the variables.

(1) When the quotients (fractions) of exponential expressions have the same base :

To simplify the rational expression, reduce the fraction to its lowest term by the properties of the exponents. Then, reduce coefficients by prime factorizations.

Example

$$\frac{a^m}{a^n} = a^{m-n}, \quad a^{-m} = \frac{1}{a^m}$$

$$\frac{2x^3}{24x} = \frac{2 \cdot x^3}{2^3 \cdot 3 \cdot x} = \frac{x^2}{2^2 \cdot 3} = \frac{x^2}{12} \qquad \therefore \quad \frac{2x^3}{24x} = \frac{x^2}{12}, \quad x \neq 0$$

$$\frac{6x^2 y^4}{15x^5 y} = \frac{2 \cdot 3 \cdot x^2 \cdot y^4}{3 \cdot 5 \cdot x^5 y} = \frac{2}{5} x^{2-5} y^{4-1} = \frac{2}{5} x^{-3} y^3 = \frac{2y^3}{5x^3} \qquad \therefore \quad \frac{6x^2 y^4}{15x^5 y} = \frac{2y^3}{5x^3}, \quad y \neq 0$$

$$\frac{x^2}{x^2 y} = \frac{1}{y} \Rightarrow \text{restriction } x \neq 0 \text{ is required.}$$

However, $y \neq 0$ is not necessary
($\because$ if $y = 0$, then both of the expressions are undefined.)

(2) When the polynomials can be factorized :

To simplify the rational expression, factor the numerator and the denominator. If the expression contains common factors, eliminate the common factors to reduce the fraction.

Example 1

$$\frac{2x^2 - 8}{x+2} = \frac{2(x^2 - 4)}{x+2} = \frac{2(x+2)(x-2)}{x+2} = 2(x-2)$$

Therefore, $\dfrac{2x^2 - 8}{x+2} = 2(x-2), \quad x \neq -2$

$$x^2 - (a+b)x + ab = (x-a)(x-b)$$

Example 2

$$\frac{2x+4}{6x^2 - 18x - 60} = \frac{2(x+2)}{6(x^2 - 3x - 10)} = \frac{2(x+2)}{2 \cdot 3(x-5)(x+2)} = \frac{1}{3(x-5)}$$

Therefore, $\dfrac{2x+4}{6x^2 - 18x - 60} = \dfrac{1}{3(x-5)}, \quad x \neq -2$

Example 3

$$\frac{2x^2 - 2}{6x^2 + 24x - 30} = \frac{2(x^2 - 1)}{6(x^2 + 4x - 5)} = \frac{2(x+1)(x-1)}{2 \cdot 3(x+5)(x-1)} = \frac{x+1}{3(x+5)}$$

Therefore, $\dfrac{2x^2 - 2}{6x^2 + 24x - 30} = \dfrac{x+1}{3(x+5)}, \quad x \neq 1$

3. Operations of Rational Expressions

(1) Adding and Subtracting Rational Expressions

For any polynomials $P(x)$, $Q(x)$, $R(x)(\neq 0)$, and $S(x)(\neq 0)$,

1) If the denominators are the same :

Keep the denominator and add or subtract the numerators

$$\frac{P(x)}{R(x)} + \frac{Q(x)}{R(x)} = \frac{P(x)+Q(x)}{R(x)} \ , \quad \frac{P(x)}{R(x)} - \frac{Q(x)}{R(x)} = \frac{P(x)-Q(x)}{R(x)}$$

Example

$$\frac{3x}{x-2} + \frac{5}{x-2} = \frac{3x+5}{x-2} \ , x \neq 2$$

$$\frac{x+1}{2x-1} - \frac{3x+5}{2x-1} = \frac{(x+1)-(3x+5)}{2x-1} = \frac{-2x-4}{2x-1} = \frac{-2(x+2)}{2x-1} \ , x \neq \frac{1}{2}$$

2) If the denominators are different :

Identify the least common denominator (LCD) of the fractions and simplify the expression by adding or subtracting the numerators, then dividing by the least common denominator.

$$\frac{P(x)}{R(x)} + \frac{Q(x)}{S(x)} = \frac{P(x)\cdot S(x)}{R(x)\cdot S(x)} + \frac{R(x)\cdot Q(x)}{R(x)\cdot S(x)} = \frac{P(x)\cdot S(x)+R(x)\cdot Q(x)}{R(x)\cdot S(x)} \ ,$$

$$\frac{P(x)}{R(x)} - \frac{Q(x)}{S(x)} = \frac{P(x)\cdot S(x)}{R(x)\cdot S(x)} - \frac{R(x)\cdot Q(x)}{R(x)\cdot S(x)} = \frac{P(x)\cdot S(x)-R(x)\cdot Q(x)}{R(x)\cdot S(x)}$$

> The least common denominator (LCD) is the least common multiple (LCM) of denominators .

Example 1

$$\frac{x-1}{2x(x+1)} + \frac{2}{3x(x-1)} = \frac{3\cdot(x-1)(x-1)}{6x(x+1)(x-1)} + \frac{2\cdot2(x+1)}{6x(x-1)(x+1)}$$

$$= \frac{3(x^2-2x+1)+4(x+1)}{6x(x+1)(x-1)} = \frac{3x^2-2x+7}{6x(x+1)(x-1)}$$

Example 2

$$\frac{3}{x+2} - \frac{2}{x-4} = \frac{3(x-4)}{(x+2)(x-4)} - \frac{2(x+2)}{(x-4)(x+2)} = \frac{3(x-4)-2(x+2)}{(x+2)(x-4)} = \frac{x-16}{(x+2)(x-4)}$$

(2) Multiplying and Dividing Rational Expressions

To multiply and divide rational expressions, common denominators are not necessary.

1) To multiply rational expressions,

Multiply the numerators, then multiply the denominators. If possible, simplify the expressions before multiplying.

Example

$$\frac{x+1}{3x} \cdot \frac{4x^2}{2(x+1)} = \frac{4(x+1)x^2}{3x \cdot 2(x+1)} = \frac{4 \cdot (x+1) \cdot x \cdot x}{3 \cdot x \cdot 2 \cdot (x+1)} = \frac{2x}{3}$$

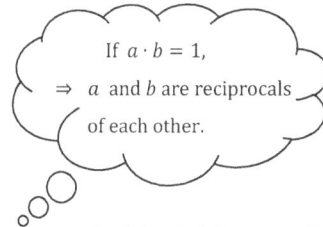

If $a \cdot b = 1$,
$\Rightarrow$ a and b are reciprocals of each other.

2) To divide rational expressions,

Convert the division into multiplication using the reciprocal of the 2nd fraction (the fraction we are dividing by). If possible, simplify the expressions before multiplying.

Example

$$\frac{2x}{(x+1)^2} \div \frac{1}{x^2-2x-3} = \frac{2x}{(x+1)^2} \cdot \frac{x^2-2x-3}{1} = \frac{2x(x^2-2x-3)}{(x+1)^2}$$

$$= \frac{2x(x-3)(x+1)}{(x+1)(x+1)} = \frac{2x(x-3)}{(x+1)}$$

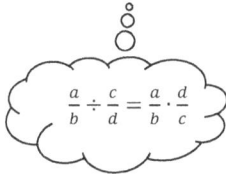

$$\frac{a}{b} \div \frac{c}{d} = \frac{a}{b} \cdot \frac{d}{c}$$

3) To divide long rational expressions (Long division of polynomials),

Let $P(x)$ and $D(x)$ be polynomials of degrees n and r, respectively, where $n \geq r$.

There exist polynomials $Q(x)$, called the *quotient*, and $R(x)$, called the *remainder*, such that :

① $P(x) = D(x) \cdot Q(x) + R(x)$ for all x

② The degree of $R(x)$ is less than the degree of $D(x)$.

Example

$\dfrac{x^3 + 2x^2 + 4x + 5}{x+2}$

$$\Rightarrow \quad \begin{array}{r} x^2 \qquad\quad + 4 \\ \hline x+2\,\overline{)\,x^3 + 2x^2 + 4x + 5} \\ x^3 + 2x^2 \qquad\qquad \\ \hline 4x + 5 \\ 4x + 8 \\ \hline -3 \end{array}$$

$\Rightarrow P(x) = D(x)Q(x) + R(x)$

$\Rightarrow \dfrac{P(x)}{D(x)} = Q(x) + \dfrac{R(x)}{D(x)}$

$Q(x) \longleftarrow$ quotient

$\overline{D(x)\,)\,P(x)} \leftarrow$ dividend

divisor

$\overline{R(x)} \leftarrow$ remainder

$\therefore \quad \dfrac{x^3 + 2x^2 + 4x + 5}{x+2} = x^2 + 4 - \dfrac{3}{x+2}$

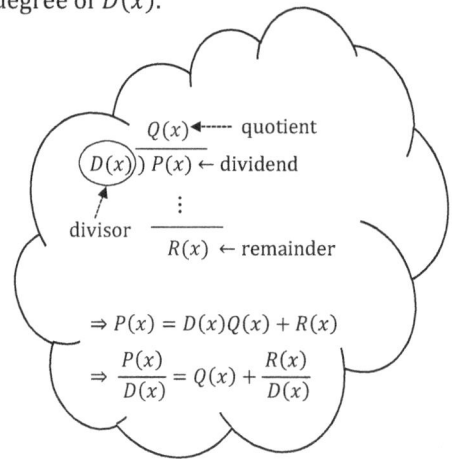

Note : If terms are missing from the dividend or the divisor, supply the missing terms with zero coefficients.

Example

$\dfrac{4x^3 + 6x^2 - 3}{2x - 1}$

$$\Rightarrow \quad \dfrac{4x^3 + 6x^2 + 0 \cdot x - 3}{2x - 1}$$

$$\Rightarrow \quad \begin{array}{r} 2x^2 + 4x\ + 2 \\ \hline 2x-1\,\overline{)\,4x^3 + 6x^2 + 0 \cdot x - 3} \\ 4x^3 - 2x^2 \qquad\qquad\quad \\ \hline 8x^2 + 0 \cdot x - 3 \\ 8x^2 - 4x \qquad\quad \\ \hline 4x\ - 3 \\ 4x\ - 2 \\ \hline -1 \end{array}$$

$\therefore \quad \dfrac{4x^3 + 6x^2 - 3}{2x - 1} = 2x^2 + 4x + 2 - \dfrac{1}{2x - 1}$

4) To simplify complex fractions,

Convert the complex fraction into a rational quotient. Rewrite the rational quotient as a product and simplify the expressions.

Example

$$\frac{\frac{x(x+1)}{x^2}}{\frac{x^2+6x+5}{x-2}} = \frac{x(x+1)}{x^2} \div \frac{x^2+6x+5}{x-2} = \frac{x(x+1)}{x^2} \cdot \frac{x-2}{x^2+6x+5} = \frac{x \cdot (x+1) \cdot (x-2)}{x \cdot x \cdot (x+5) \cdot (x+1)}$$

$$= \frac{x-2}{x(x+5)} \quad , x \neq 0, \ x \neq -1$$

$$\frac{\frac{a}{b}}{\frac{c}{d}} = \frac{ad}{bc}$$

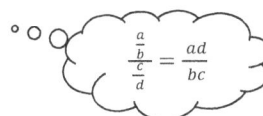

5) Synthetic Division

When the divisor is a linear binomial (: x + constant), the long division of polynomials is solved by synthetic division.

Example Solve $\dfrac{x^3+2x^2+4x+5}{x+2}$.

Write the coefficients (1, 2, 4, 5) of the dividend, $x^3 + 2x^2 + 4x + 5$, in a line

$$\begin{array}{|c} 1 \ \ 2 \ \ 4 \ \ 5 \\ \hline \end{array}$$

To the left side of the coefficients, write the opposite (-2) of the constant term of the divisor, $x + 2$

$$-2 \begin{array}{|ccc} 1 & 2 & 4 & 5 \\ \hline \end{array}$$

Bring first coefficient (1) below the line

$$-2 \begin{array}{|ccc} 1 & 2 & 4 & 5 \\ \hline 1 \end{array}$$

Multiply the left number (-2) by the number (1) below the line and place the product $(-2 \cdot 1 = -2)$ under the second coefficient (2)

$$-2 \begin{array}{|cccc} 1 & 2 & 4 & 5 \\ & -2 \\ \hline 1 \end{array}$$

Add the second coefficient (2) with the product (-2). Write the sum below the line to complete the column

$$-2 \begin{array}{|cccc} 1 & 2 & 4 & 5 \\ & -2 \\ \hline 1 & 0 \end{array}$$

Multiply the left number (-2) by the number (0) and place the product ($-2 \cdot 0 = 0$) under the third coefficient (4). Add the third coefficient (4) and the product (0). Write the sum below the line to complete the column

$$
\begin{array}{r|rrrr}
-2 & 1 & 2 & 4 & 5 \\
 & & -2 & 0 & \\
\hline
 & 1 & 0 & 4 &
\end{array}
$$

Repeat the steps, multiply, place, and add.

$$
\begin{array}{r|rrrr}
-2 & 1 & 2 & 4 & 5 \\
 & & -2 & 0 & -8 \\
\hline
 & 1 & 0 & 4 & -3
\end{array}
$$

All the numbers below the line except the last number (-3) represent the coefficients of the quotient. The last number (-3) represents the remainder. Since we divided the dividend by a linear binomial which has a power of 1, the power of the quotient is 1 less than the power of the dividend. Since the dividend has a power of 3, the quotient must have a power of 2. Therefore, the quotient is $1 \cdot x^2 + 0 \cdot x + 4$ and remainder is -3. That is,

$$
\frac{x^3+2x^2+4x+5}{x+2} = x^2 + 4 - \frac{3}{x+2}
$$

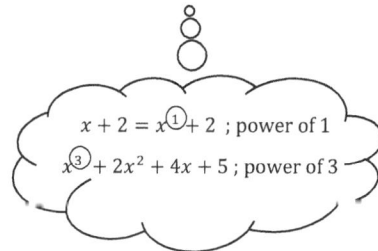

$x + 2 = x^① + 2$; power of 1
$x^③ + 2x^2 + 4x + 5$; power of 3

Example

Solve $\dfrac{4x^3+6x^2-3}{2x-1}$.

Since the divisor is $2x - 1$, express it as the form of $x +$ constant (: linear binomial).

Dividing the expression by 2, we get $\dfrac{4x^3+6x^2-3}{2x-1} = \dfrac{2x^3+3x^2-\frac{3}{2}}{x-\frac{1}{2}}$.

Now we can use synthetic division.

Since x-term is missing, supply the x-term with a zero coefficient.

$$
\begin{array}{r|rrrr}
\frac{1}{2} & 2 & 3 & 0 & -\frac{3}{2} \\
 & & 1 & 2 & 1 \\
\hline
 & 2 & 4 & 2 & -\frac{1}{2}
\end{array}
$$

$\therefore \dfrac{2x^3+3x^2-\frac{3}{2}}{x-\frac{1}{2}} = 2x^2 + 4x + 2 - \dfrac{\frac{1}{2}}{x-\frac{1}{2}} = 2x^2 + 4x + 2 - \dfrac{1}{2x-1}$

Therefore, $\dfrac{4x^3+6x^2-3}{2x-1} = 2x^2 + 4x + 2 - \dfrac{1}{2x-1}$

8-2 Solving Rational Equations and Inequalities

1. Rational Equations

(1) Using the Cross Product (Multiplication)

In a proportion,

$$\frac{P(x)}{R(x)} = \frac{Q(x)}{S(x)} \quad \Rightarrow \quad P(x)S(x) = Q(x)R(x), \text{ where } R(x) \neq 0, S(x) \neq 0$$

A proportion is an equality of two ratios.

Example

$$\frac{20}{x^2+4x+3} = \frac{x}{x+3}$$

$$\Rightarrow \frac{20}{(x+1)(x+3)} = \frac{x}{x+3}$$

$$\Rightarrow \quad x(x+1)(x+3) = 20(x+3) \quad \Rightarrow \quad x(x+1) = 20$$

$$\Rightarrow \quad x^2 + x - 20 = 0 \quad \Rightarrow \quad (x+5)(x-4) = 0 \quad \Rightarrow \quad x = -5 \text{ or } x = 4$$

∴ The solution to the equation is $x = -5$ or $x = 4$

(2) Using the Least Common Denominator (LCD)

If a rational equation consists of two or more fractions on one side of the equal sign, combine the expressions on the side containing fractions by using their least common denominator (LCD). Then simplify the expression.

Example

$$3 + \frac{3}{x-2} = -2x$$

$$\Rightarrow \frac{3(x-2)}{x-2} + \frac{3}{x-2} = -2x \quad \text{the LCD of } \frac{3}{1} \text{ and } \frac{3}{x-2} \text{ is } x-2$$

$$\Rightarrow \frac{3x-3}{x-2} = -2x \quad \text{combine the expressions on the left side of the equation}$$

$$\Rightarrow -2x(x-2) = 3x - 3 \quad \text{cross multiply to eliminate a fraction}$$

$$\Rightarrow 2x^2 - x - 3 = 0$$

$$\Rightarrow (2x-3)(x+1) = 0 \quad \text{factor the quadratic equation}$$

$$\Rightarrow 2x - 3 = 0 \text{ or } x + 1 = 0$$

$$\Rightarrow x = \frac{3}{2} \text{ or } x = -1$$

Therefore, the solution to the equation is $x = \frac{3}{2}$ or $x = -1$.

2. Rational Inequalities

(1) Critical Numbers

If the denominator of a rational expression is equal to zero, the expression is not defined. Any values by which the expression is undefined or equal to zero are called the *critical numbers* of the rational expression. These critical numbers divide a number line into intervals.

(2) Solving Rational Inequalities

To solve rational inequalities, choose a test value from each interval to identify the solutions.

Example 1

$$\frac{x-1}{x+2} > 0$$

If the denominator equals zero $(x + 2 = 0)$, then $x = -2$, and the expression is undefined.

If the numerator equals zero $(x - 1 = 0)$, then $x = 1$, and the expression is equal to zero.

Thus, the critical numbers of the rational expression are $x = -2$ and $x = 1$.

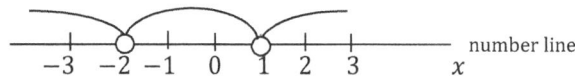

These critical numbers divide the number line into three intervals.

Case 1. $x < -2$

⇒ Choose a test value, such as $x = -3$, and substitute the test value into the expression. Then, $\frac{x-1}{x+2} = \frac{-3-1}{-3+2} = \frac{-4}{-1} = 4$

Since $4 > 0$, $\frac{x-1}{x+2} > 0$ is true.

Case 2. $-2 < x < 1$

⇒ Choose a test value, such as $x = 0$, and substitute the test value into the expression. Then, $\frac{x-1}{x+2} = \frac{0-1}{0+2} = -\frac{1}{2}$

Since $-\frac{1}{2} < 0$, $\frac{x-1}{x+2} > 0$ is false.

Case 3. $x > 1$

⇒ Choose a test value, such as $x = 2$, and substitute the test value into the expression. Then, $\frac{x-1}{x+2} = \frac{2-1}{2+2} = \frac{1}{4}$

Since $\frac{1}{4} > 0$, $\frac{x-1}{x+2} > 0$ is true.

Therefore, the solution to the rational inequality is $x < -2$ or $x > 1$.

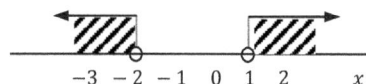

Note 1. If $\dfrac{P(x)}{Q(x)} > 0$ *or* $\dfrac{P(x)}{Q(x)} < 0$, *then the critical numbers are open points on the number line.*

Note 2. If $\dfrac{P(x)}{Q(x)} \geq 0$ *or* $\dfrac{P(x)}{Q(x)} \leq 0$, *then the critical numbers are closed points on the number line.*

However, the number for which the denominator equals zero (which makes the expression undefined) is always an open point on a number line.

Example 2

$$\frac{x}{3} \leq \frac{5}{x+2}$$

$$\Rightarrow \frac{x}{3} - \frac{5}{x+2} \leq 0$$

$$\Rightarrow \frac{x(x+2)-15}{3(x+2)} \leq 0$$

$$\Rightarrow \frac{x^2+2x-15}{3(x+2)} \leq 0$$

$$\Rightarrow \frac{(x+5)(x-3)}{3(x+2)} \leq 0$$

When $a > b$,

① If $(x-a)(x-b) > 0$ (positive)

$\Rightarrow x > a$ or $x < b$

∴ x is greater than the larger number or x is less than the smaller number

② If $(x-a)(x-b) < 0$ (negative)

$\Rightarrow b < x < a$: x is between the smaller number and the larger number

∴ $\dfrac{x-1}{x+2} > 0 \Rightarrow (x-1)(x+2) > 0$

$\Rightarrow x > 1$ or $x < -2$

The critical numbers are $x = -2, x = -5,$ and $x = 3.$

number line
x

The number line is now divided into four intervals.

If $x = -2$, then the expression is undefined. So, $x = -2$ is not included. We draw an open point for this value on the number line.

Case1. $x \leq -5 \Rightarrow$ try $x = -6$: $\dfrac{-6}{3} \leq \dfrac{5}{-6+2}$; $-2 \leq \dfrac{5}{-4}$ (true)

Case2. $-5 \leq x < -2 \Rightarrow$ try $x = -3$: $\dfrac{-3}{3} \leq \dfrac{5}{-3+2}$; $-1 \leq -5$ (false)

Case3. $-2 < x \leq 3 \Rightarrow$ try $x = 0$: $\dfrac{0}{3} \leq \dfrac{5}{0+2}$; $0 \leq \dfrac{5}{2}$ (true)

Case4. $x \geq 3 \Rightarrow$ try $x = 6$: $\dfrac{6}{3} \leq \dfrac{5}{6+2}$; $2 \leq \dfrac{5}{8}$ (false)

Therefore, the solution to the rational inequality is $x \leq -5$ or $-2 < x \leq 3$.

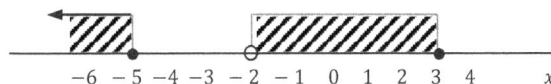

Exercises

#1. Simplify each expression.

(1) $\dfrac{2x^5}{x^2}$

(2) $\dfrac{12x^4y}{56x^2y^6}$

(3) $\dfrac{8-2x}{3x-12}$

(4) $\dfrac{x^2+6x+8}{x^2+3x-4}$

(5) $\dfrac{3}{x} + \dfrac{1}{2x}$

(6) $\dfrac{2x}{x^2-4} - \dfrac{3}{x+2}$

(7) $\dfrac{x+1}{x^2-9} - \dfrac{x}{x^2-2x-15}$

(8) $\dfrac{x+3}{x^2-4} \cdot \dfrac{x+2}{x^2+2x-3}$

(9) $\dfrac{x^2-16}{(x+5)^2} \div \dfrac{x+4}{x^2+2x-15}$

(10) $\dfrac{\frac{2}{x^3}}{\frac{8}{x^5}}$

(11) $\dfrac{\frac{2x^2-5x-3}{x-3}}{\frac{2x^2-7x-4}{2x}}$

(3) $(3x^3 + 2x^2 + 1) \div (x - 2)$

#2. Divide the following long rational expressions

(1) $(x^2 + 5x - 6) \div (x - 2)$

#3. Solve the following rational equations

(1) $\dfrac{3x}{2} = \dfrac{5}{8}$

(2) $\dfrac{2}{3x-4} = \dfrac{-1}{2-4x}$

(3) $7 + \dfrac{4}{x-1} = -2x$

(2) $(2x^2 - 20) \div (x + 3)$

(4) $\dfrac{1}{x+2} + \dfrac{2}{x^2-4} = \dfrac{4}{x-2}$

#4. Solve the following inequalities

(1) $\dfrac{x-4}{x+3} > 0$

(2) $\dfrac{x+5}{x-2} \leq 0$

(3) $\dfrac{x}{2} \geq \dfrac{5}{x-3}$

(4) $\dfrac{2x^2-5x-3}{x-4} > 0$

Solutions Manual

Algebra

Part II. Expressions

Solutions for Chapter 1

 # 1 Find the value for each expression.

(1) $x + 3$ if $x = -3$; $x + 3 = -3 + 3 = 0$

(2) $-(x + 1)$ if $x = -5$; $-(x + 1) = -(-5 + 1) = -(-4) = 4$

(3) $x^3 - 2x - 5$ if $x = -1$; $x^3 - 2x - 5 = (-1)^3 - 2(-1) - 5 = -1 + 2 - 5 = -4$

(4) $2xy - 4$ if $x = -3, y = 2$; $2xy - 4 = 2(-3)(2) - 4 = -12 - 4 = -16$

(5) $x^2 - 5y$ if $x = -2, y = -3$; $x^2 - 5y = (-2)^2 - 5(-3) = 4 + 15 = 19$

(6) $\frac{2}{x} + \frac{3}{y}$ if $x = -\frac{1}{4}, y = -\frac{1}{6}$; $\frac{2}{x} + \frac{3}{y} = 2(-4) + 3(-6) = -8 - 18 = -26$

(7) $\frac{y}{x} - \frac{x}{y}$ if $x = 2, y = -3$

$\frac{y}{x} - \frac{x}{y} = y\left(\frac{1}{2}\right) - x\left(-\frac{1}{3}\right) = (-3)\left(\frac{1}{2}\right) - (2)\left(-\frac{1}{3}\right) = -\frac{3}{2} + \frac{2}{3} = -\frac{9}{6} + \frac{4}{6} = -\frac{5}{6}$

(8) $x^{99} - x^6$ if $x = -1$; $x^{99} - x^6 = (-1)^{99} - (-1)^6 = (-1) - (1) = -2$

(9) $\frac{3}{a} - 2b^2$ if $a = \frac{1}{4}, b = -3$; $\frac{3}{a} - 2b^2 = 3(4) - 2(-3)^2 = 12 - 2(9) = 12 - 18 = -6$

(10) $\frac{1}{a} - \frac{2}{b} - \frac{3}{c}$ if $a = -\frac{1}{4}, b = -\frac{1}{2}, c = -6$

$\frac{1}{a} - \frac{2}{b} - \frac{3}{c} = 1(-4) - 2(-2) - 3\left(-\frac{1}{6}\right) = -4 + 4 + \frac{1}{2} = \frac{1}{2}$

(11) $2(a - b) - (a^2 - b^2)$ if $a = -2, b = 3$

$2(a - b) - (a^2 - b^2) = 2(-2 - 3) - (4 - 9) = 2(-5) - (-5) = -10 + 5 = -5$

(12) $|a - 2| - |3ab - a|$ if $a = -1, b = 2$

$|a - 2| - |3ab - a| = |-1 - 2| - |3(-1)(2) - (-1)| = |-3| - |-6 + 1| = |-3| - |-5|$

$= 3 - (5) = -2$

 # 2 Simplify each expression.

(1) $\frac{1}{2}x \cdot (-6) = \frac{1}{2}(-6)x = -3x$

(2) $-\frac{2}{3}(6x - 9) = -\frac{2}{3}(6x) + \frac{2}{3}(9) = -4x + 6$

(3) $(6x - 2) \div \left(-\frac{3}{10}\right) = (6x - 2) \times \left(-\frac{10}{3}\right) = 6x\left(-\frac{10}{3}\right) + 2\left(\frac{10}{3}\right) = -20x + \frac{20}{3}$

(4) $\frac{1}{2}(4x - 6) - \frac{2}{3}\left(9x - \frac{3}{4}\right) = \left(\frac{1}{2}(4x) - \frac{2}{3}(9x)\right) - \frac{1}{2}(6) + \frac{2}{3}\left(\frac{3}{4}\right) = (2x - 6x) - 3 + \frac{1}{2} = -4x - 2\frac{1}{2}$

(5) $-3x + 6x - 2x - 5 = (-3x + 6x - 2x) - 5 = x - 5$

(6) $2x + 4y - \{3x - (5 - 2y)\} - 3 = 2x + 4y - 3x + (5 - 2y) - 3 = -x + 2y + 2$

(7) $(2a - 3b) - (5a - 2b) - (4 - a) = 2a - 3b - 5a + 2b - 4 + a = -2a - b - 4$

(8) $\quad 3m^2 - (5m - m^2 - 1) + 2m = 3m^2 - 5m + m^2 + 1 + 2m = 4m^2 - 3m + 1$

(9) $\quad \dfrac{3t^3 - 4t^2}{2t} = \dfrac{3t^3}{2t} - \dfrac{4t^2}{2t} = \dfrac{3}{2}t^2 - 2t$

(10) $\quad \dfrac{2a^2b - 3ab^2 - 5ab}{ab} = \dfrac{2a^2b}{ab} - \dfrac{3ab^2}{ab} - \dfrac{5ab}{ab} = 2a - 3b - 5$

(11) $\quad (3x - 9) \div \dfrac{3}{2} - 8\left(\dfrac{3}{4}x - 2\right) = (3x - 9) \times \dfrac{2}{3} - 8\left(\dfrac{3}{4}x - 2\right) = (2x - 6) - (6x - 16) = -4x + 10$

(12) $\quad \dfrac{x-2}{3} - \dfrac{2x-1}{4} - \dfrac{3-x}{2} = \left(\dfrac{x}{3} - \dfrac{2x}{4} + \dfrac{x}{2}\right) - \dfrac{2}{3} + \dfrac{1}{4} - \dfrac{3}{2} = \dfrac{4x - 6x + 6x}{12} + \dfrac{-8 + 3 - 18}{12} = \dfrac{1}{3}x - 1\dfrac{11}{12}$

♯ 3 Find an expression for the perimeter.

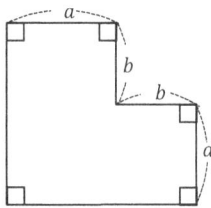

$$a + b + b + a + (a + b) + (b + a) = 4a + 4b = 4(a + b)$$

♯ 4 Find an expression for the shaded area.

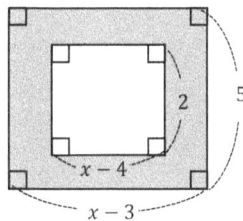

$$5(x - 3) - 2(x - 4) = 5x - 15 - 2x + 8 = 3x - 7$$

♯ 5 Which expression is different from the others? (4)

(1) $x \div (y \times z) = x \div yz = x \times \dfrac{1}{yz} = \dfrac{x}{yz}$

(2) $x \div y \div z = x \times \dfrac{1}{y} \times \dfrac{1}{z} = \dfrac{x}{yz}$

(3) $x \times \dfrac{1}{y} \div z = x \times \dfrac{1}{y} \times \dfrac{1}{z} = \dfrac{x}{yz}$

(4) $x \div (y \div z) = x \div \left(y \times \dfrac{1}{z}\right) = x \div \left(\dfrac{y}{z}\right) = x \times \dfrac{z}{y} = \dfrac{xz}{y}$

(5) $x \times \dfrac{1}{y} \times \dfrac{1}{z} = \dfrac{x}{yz}$

 # 6 Find $a + b$ and $a - b$ when a is a coefficient of x, and b is a constant for the expression :

$\dfrac{4x-3}{2} - \dfrac{2x-1}{3}$.

$\dfrac{4x-3}{2} - \dfrac{2x-1}{3} = \dfrac{12x-9-4x+2}{6} = \dfrac{8x-7}{6} = ax + b$

$\therefore\ a = \dfrac{8}{6},\ b = -\dfrac{7}{6}$

Therefore, $a + b = \dfrac{1}{6}$ and $a - b = \dfrac{15}{6} = \dfrac{5}{2}$

 # 7 The coefficient of x is 3 and the constant is 5 for the form $2x + b - (ax + 3)$. Find $a \cdot b$ and $\dfrac{a}{b}$

Since $2x + b - (ax + 3) = (2 - a)x + (b - 3)$, $2 - a = 3$, $b - 3 = 5$.

$\therefore\ a = -1,\ b = 8$

Therefore, $a \cdot b = -8$ and $\dfrac{a}{b} = -\dfrac{1}{8}$

 # 8 For two expressions A and B, if you add $2x + 3$ to A then you get $5x + 7$ and if you subtract $-3x - 4$ from B then you get $2x + 5$. What is $2A - 3B$?

Since $A + (2x + 3) = 5x + 7$ and $B - (-3x - 4) = 2x + 5$,

$A = 3x + 4$ and $B = -x + 1$.

$\therefore\ 2A - 3B = 9x + 5$

 # 9 Solve the following equations for x :

(1) $3x - 2 = 7$; $3x = 9$; $x = 3$

(2) $2x + 3 = 3x - 2$; $2x - 3x = -2 - 3$; $-x = -5$; $x = 5$

(3) $5x - 2 = \dfrac{1}{2}x - 1\dfrac{1}{4}$; $5x - \dfrac{1}{2}x = -1\dfrac{1}{4} + 2$; $\dfrac{9}{2}x = \dfrac{3}{4}$; $\dfrac{18}{4}x = \dfrac{3}{4}$; $18x = 3$; $x = \dfrac{1}{6}$

(4) $0.2x - 0.3 = 0.4x - 0.5$; $0.2x - 0.4x = -0.5 + 0.3$; $-0.2x = -0.2$; $x = 1$

(5) $\dfrac{3}{4}\left(x - \dfrac{1}{3}\right) = \dfrac{1}{2}\left(\dfrac{1}{5} + 4x\right)$; $\dfrac{3}{4}x - \dfrac{1}{4} = \dfrac{1}{10} + 2x$; $\dfrac{3}{4}x - 2x = \dfrac{1}{10} + \dfrac{1}{4}$; $-\dfrac{5}{4}x = \dfrac{7}{20}$; $x = -\dfrac{7}{25}$

(6) $\dfrac{x-3}{2} - 1 = \dfrac{x}{4} - 3$; $2(x - 3) - 4 = x - 12$; $2x - x = -12 + 6 + 4$; $x = -2$

(7) $3(1 - 2x) + 7 = -2x - 2$; $3 - 6x + 7 = -2x - 2$; $-6x + 2x = -2 - 10$; $-4x = -12$; $x = 3$

(8) $3 - \dfrac{2x-1}{3} = 5x - \dfrac{x-2}{6}$; $18 - 2(2x - 1) = 30x - (x - 2)$; $-4x - 29x = -18$; $-33x = -18$; $x = \dfrac{6}{11}$

10 Find a constant a, that makes the equation $4(a + x) = 2(2x + 3) + 6$ true for all values of x.

$$4(a + x) = 2(2x + 3) + 6$$

$$\Rightarrow 4a + 4x = 4x + 6 + 6 \qquad \Rightarrow 4a + 4x = 4x + 12$$

$$\Rightarrow 4a = 12 \qquad \Rightarrow a = 3$$

11 For any positive integers a, b, and c, a is divided by b and the remainder is c. Express the quotient using a, b, and c.

Let Q be the quotient. Since $a = bQ + c$, $a - c = bQ$. Therefore, the quotient is $\frac{a-c}{b}$.

12 For any non-zero constants a, $b(b \neq 1)$, and c,

find the value of $\frac{1}{abc}$ such that $a + \frac{1}{b} = 1$ and $b + \frac{1}{c} = 1$.

$$a + \frac{1}{b} = 1 \Rightarrow a = 1 - \frac{1}{b} = \frac{b-1}{b}$$

$$b + \frac{1}{c} = 1 \Rightarrow \frac{1}{c} = 1 - b$$

$$\therefore \frac{1}{abc} = \frac{1}{a} \cdot \frac{1}{b} \cdot \frac{1}{c} = \frac{b}{b-1} \cdot \frac{1}{b} \cdot \frac{1-b}{1} = -1$$

13 $a = \frac{2}{3}, b = \frac{3}{4}$, and $c = -\frac{4}{5}$. Find the value of $\frac{ab+bc+ca}{abc}$.

$$\frac{ab+bc+ca}{abc} = \frac{ab}{abc} + \frac{bc}{abc} + \frac{ca}{abc} = \frac{1}{c} + \frac{1}{a} + \frac{1}{b} = -\frac{5}{4} + \frac{3}{2} + \frac{4}{3} = \frac{-15+18+16}{12} = \frac{19}{12} = 1\frac{7}{12}$$

14 Find the sum of all possible solutions for the equation $|2x - 3| = 5$.

If $2x - 3 \geq 0 \left(x \geq \frac{3}{2} \right)$, then $|2x - 3| = 2x - 3 = 5$; $2x = 8$; $x = 4$

If $2x - 3 < 0 \left(x < \frac{3}{2} \right)$, then $|2x - 3| = -(2x - 3) = 5$; $2x = -2$; $x = -1$

$\therefore$ The sum of all possible solutions is $4 + (-1) = 3$.

 # 15 $A = 2x - 3$, $B = 3x + 4$. **The ratio of A to B is $3 : 5$.**

When a is the solution of x, find the value of $-\frac{a}{3} + 3$.

Since the ratio of A to B is $3 : 5$,

$$2x - 3 : 3x + 4 = 3 : 5 \quad ; \quad \frac{2x-3}{3x+4} = \frac{3}{5} \quad ; \quad 5(2x - 3) = 3(3x + 4) \quad ; \quad x = 27$$

Since a is the solution of x, $\quad -\frac{a}{3} + 3 = -\frac{27}{3} + 3 = -9 + 3 = -6$

 # 16 **For any constants a, b, the solution of the equation $3x - 2 = ax - 4$ is $x = -1$ and the solution**

of the equation $\frac{1}{2}x + b = ax + 3$ is $x = -2$. Find $a \cdot b$

$$3x - 2 = ax - 4 \ \Rightarrow \ 3(-1) - 2 = a(-1) - 4 \ \Rightarrow \ -5 = -a - 4 \Rightarrow \ a = 1$$

$$\frac{1}{2}x + b = ax + 3 \ \Rightarrow \ \frac{1}{2}(-2) + b = a(-2) + 3 \ \Rightarrow 2a + b = 4 \ \Rightarrow \ 2(1) + b = 4 \ \Rightarrow \ b = 2$$

$$\therefore \ a \cdot b = 1 \cdot 2 = 2$$

 # 17 **The solution of the equation $2ax + 5 = -3$ is half of the solution of the equation $x - 5 = 3x + 7$.**

Find the value of $3a - 4$.

$$x - 5 = 3x + 7 \Rightarrow \ 2x = -12 \ \Rightarrow \ x = -6$$

So, the solution x of an equation $2ax + 5 = -3$ is $x = -3$.

So, $2a(-3) + 5 = -3 \ \Rightarrow \ -6a + 5 = -3 \ \Rightarrow \ -6a = -8 \ \Rightarrow \ a = \frac{4}{3}$

Therefore, $3a - 4 = 3 \cdot \frac{4}{3} - 4 = 0$.

 # 18 **For any constants a and b, $\frac{1}{a} - \frac{1}{b} = 3 \ (ab \neq 0)$. Find the value of $\frac{5a-3ab-5b}{a-b}$.**

$$\frac{1}{a} - \frac{1}{b} = 3 \ \Rightarrow \ \frac{b-a}{ab} = 3 \ \Rightarrow \ b - a = 3ab \Rightarrow a - b = -3ab$$

$$\therefore \ \frac{5a-3ab-5b}{a-b} = \frac{5(a-b)-3ab}{a-b} = \frac{5(-3ab)-3ab}{-3ab} = \frac{-18ab}{-3ab} = 6$$

19 $\begin{cases} (1) \ \dfrac{a+3}{4} - \dfrac{2x-2}{3} = 1 \\ (2) \ \dfrac{3a-2}{2} - \dfrac{2a-x}{3} = 1 \end{cases}$

When the ratio of the solution of (1) to the solution of (2) is $1 : 4$, find the value of a.

(1) $\dfrac{a+3}{4} - \dfrac{2x-2}{3} = 1$

$\Rightarrow \dfrac{3a+9-8x+8}{12} = 1 \Rightarrow 3a + 17 - 8x = 12 \Rightarrow 8x = 3a + 5 \Rightarrow x = \dfrac{3a+5}{8}$.

(2) $\dfrac{3a-2}{2} - \dfrac{2a-x}{3} = 1$

$\Rightarrow \dfrac{9a-6-4a+2x}{6} = 1 \Rightarrow 5a - 6 + 2x = 6 \Rightarrow 2x = 12 - 5a \Rightarrow x = \dfrac{12-5a}{2}$

$\therefore \ \dfrac{3a+5}{8} : \dfrac{12-5a}{2} = 1 : 4 \ ; \ 1 \cdot \dfrac{12-5a}{2} = 4 \cdot \dfrac{3a+5}{8} \ ; \ \dfrac{12-5a}{2} = \dfrac{3a+5}{2} \ ; \ 12 - 5a = 3a + 5 \ ; \ 8a = 7$

Therefore, $a = \dfrac{7}{8}$.

20 **For any x, the equation $3x - 5a = 2bx + 6$, where a and b are constants, is always true.**

Find the value of $\dfrac{a}{2b}$.

$3x - 5a = 2bx + 6 \Rightarrow (3 - 2b)x = 5a + 6 \Rightarrow (3 - 2b)x + 0 = 0 \cdot x + 5a + 6$

$\therefore \ 3 - 2b = 0, \ 5a + 6 = 0 \quad \therefore \ b = \dfrac{3}{2} \ , \ a = -\dfrac{6}{5}$

Therefore, $\dfrac{a}{2b} = \dfrac{-\dfrac{6}{5}}{2 \cdot \dfrac{3}{2}} = -\dfrac{6}{15} = -\dfrac{2}{5}$

21 **The solution of an equation $\dfrac{2x-5a}{3} + x + 4 = 8$ is a negative integer. Find the greatest value of a.**

$\dfrac{2x-5a}{3} + x + 4 = 8$

$\Rightarrow 2x - 5a + 3x + 12 = 24 \Rightarrow 5x - 5a = 12 \Rightarrow 5x = 5a + 12 \Rightarrow x = \dfrac{5a+12}{5}$

Since $x = \dfrac{5a+12}{5} < 0$, $\dfrac{5a+12}{5} = -1, -2, -3, \cdots$

So, $\dfrac{5a+12}{5} = -1$ to get the greatest value of a.

$\dfrac{5a+12}{5} = -1 \Rightarrow 5a + 12 = -5 \Rightarrow 5a = -17 \Rightarrow a = -\dfrac{17}{5}$

Therefore, the greatest value of a is $a = -\dfrac{17}{5} = -3\dfrac{2}{5}$.

22 $a@b = ab^2 + a^2b$

When $\frac{1}{a} = 2$, $\frac{1}{b} = -3$, find the value of $b@a$.

$$b@a = ba^2 + b^2a = -\frac{1}{3} \cdot \frac{1}{4} + \frac{1}{9} \cdot \frac{1}{2} = -\frac{1}{12} + \frac{1}{18} = \frac{-3+2}{36} = -\frac{1}{36}$$

23 How much water should be added to 30 ounces of a 20% salt solution to produce a 15% solution?

Let x be the amount of water added. Then,

$$20\% \cdot 30 + 0\% \cdot x = 15\% \cdot (30 + x)$$

$$\Rightarrow \frac{20}{100} \cdot 30 = \frac{15}{100} \cdot (30 + x)$$

$$\Rightarrow 6 = \frac{3}{20} \cdot (30 + x)$$

$$\Rightarrow 600 = 15(30 + x)$$

$$\Rightarrow 40 = 30 + x$$

$$\Rightarrow x = 10$$

Therefore, 10 ounces of water should be added.

24 Richard has 20 ounces of a 15% of salt solution. How much salt should he add to make it a 20% solution?

Let x be the amount of salt added. Then,

$$15\% \cdot 20 + 100\% \cdot x = 20\% \cdot (20 + x)$$

$$\Rightarrow \frac{15}{100} \cdot 20 + x = \frac{20}{100} \cdot (20 + x)$$

$$\Rightarrow 3 + x = 4 + \frac{x}{5}$$

$$\Rightarrow \frac{4}{5}x = 1$$

$$\Rightarrow x = \frac{5}{4}$$

Therefore, he should add $\frac{5}{4}$ ounces of salt.

 # 25 Richard drives to place *A* at 30 miles per hour. 20 minutes after he departs, Nichole goes to the place *A* at 50 miles per hour. How long will it take until Richard meets Nichole?

Let x be the time for Nichole to meet Richard. Then,

$$50 \cdot x = 30 \cdot (x + \tfrac{20}{60})$$

$$\Rightarrow 50\,x = 30x + 10$$

$$\Rightarrow 20\,x = 10$$

$$\Rightarrow x = \tfrac{1}{2} \text{ (30 minutes)}$$

Therefore, $50\ (= 30 + 20)$ minutes will be taken.

 # 26 Richard wants to make 50 ounces of a 10% salt solution by mixing a 7% salt solution with a 15% salt solution. How many ounces of a 7% salt solution must be mixed?

Let x be the amount of a 7% salt solution. Then,

$$7\% \cdot x + 15\% \cdot (50 - x) = 10\% \cdot 50$$

$$\Rightarrow \frac{7}{100} \cdot x + \frac{15}{100} \cdot (50 - x) = \frac{10}{100} \cdot 50$$

$$\Rightarrow 7x + 750 - 15x = 500$$

$$\Rightarrow 8x = 250$$

$$\Rightarrow x = \frac{250}{8} = 31\tfrac{1}{4}$$

Therefore, $31\tfrac{1}{4}$ ounces of a 7% salt solution must be mixed.

 # 27 Richard spends two-thirds of the money in his pocket to buy a book. He now has 4 dollars left. How much money did he have at the beginning?

Let x be the total money he had at the beginning. Then,

$$x - \tfrac{2}{3}x = 4 \,;\; \tfrac{1}{3}x = 4 \,;\; x = 12$$

So, he had 12 dollars at the beginning.

 # 28 A bag is on sale for a 15% discount. Nichole paid $60, including a 6% sales tax. What was the original price of the bag (rounded to the nearest hundredth)?

Let x be the original price of the bag and A be the discounted price of the bag. Then,

$A = x - x \cdot 15\%$ and $A + A \cdot 6\% = 60$

$\Rightarrow \quad \dfrac{85}{100}x + \dfrac{85}{100}x \cdot \dfrac{6}{100} = 60$

$\Rightarrow \quad \dfrac{85}{100}x \left(1 + \dfrac{6}{100}\right) = 60$

$\Rightarrow \quad 85x \cdot \dfrac{106}{100} = 6000$

$\Rightarrow \quad 85x = 6000 \cdot \dfrac{100}{106}$

$\Rightarrow \quad x = 66.5926 \cdots$

Therefore, the original price of the bag was $ 66.59

 # 29 Richard's aunt is 51 years old. She is three times as old as the sum of the ages of Richard and his sister. Richard is 7 years younger than his sister. How old is Richard's sister?

Let R be the age of Richard and S be the age of his sister. Then, $R = S - 7$.

So, $51 = 3(S + R) = 3(S + S - 7) = 3(2S - 7) = 6S - 21$

$\therefore \ 6S = 51 + 21 = 72 \ ; \ S = 12$

Therefore, Richard's sister is 12 years old.

 # 30 The sum of three consecutive odd integers is 153. Find the biggest number of these three integers.

Let n, $n + 2$, and $n + 4$ be the three consecutive odd integers. Then,

$n + (n + 2) + (n + 4) = 153 \ ; \ 3n + 6 = 153 \ ; \ 3n = 147 \ ; \ n = 49$

So, the three consecutive odd integers are $49, 51$, and 53.

Therefore, the biggest number of the three is 53.

 # 31 **The tens digit of a certain two-digit integer is 3. If the digits of the number are interchanged, the number will be 1 less than two times the original number. Find the original number.**

Let x be the units' digit. Then, the original number is $30 + x$.

So, the integer after interchanged is $10x + 3$.

$10x + 3 = 2(30 + x) - 1$; $x = 7$

Therefore, the original number is 37 .

 # 32 **Richard is 5 years old and Nichole is 12 years old. In how many years will Nichole be two times Richard's age?**

$12 + x = 2(5 + x)$; $x = 2$

So, 2 years later.

 # 33 **Richard takes 3 hours to finish a job if he works alone. Nichole takes 2 hours to finish the same job if she works alone. How long will it take them to finish the job if they work together?**

Consider the total amount of the job to complete is 1. Let x be the amount of time to work in hours if they work together to complete the job. Then, Richard's working rate is $\frac{1}{3}$ of the job in 1 hour and Nichole's working rate is $\frac{1}{2}$ of the job in 1 hour. Therefore, Richard works $\frac{1}{3}x$ of the job in x hours and Nichole works $\frac{1}{2}x$ of the job in x hours. Since they will be working together, the sum of the two parts equals one complete the job. Therefore,

$\frac{1}{3}x + \frac{1}{2}x = 1$; $\left(\frac{1}{3} + \frac{1}{2}\right)x = 1$; $\frac{5}{6}x = 1$; $x = \frac{6}{5} = 1\frac{1}{5}$

Therefore, working together, they can complete the job in $1\frac{1}{5}$ hours (1 hour 12 minutes).

OR <u>Alternate approach</u>

In 1 hour, Richard can do $\frac{1}{3}$ of the job and Nichole can do $\frac{1}{2}$ of the job.

Together, in 1 hour, they can do $\frac{1}{3} + \frac{1}{2} = \frac{5}{6}$ of the job.

Together, the time it takes for them to complete the job is $\dfrac{1\ \text{hour}}{\frac{5}{6}\ \text{job/hour}} = \frac{6}{5}\ \text{hours} = 1\frac{1}{5}\ \text{hours}$

 # 34 Nichole took 8 days to finish a job and Richard took 6 days to finish the same job. If Nichole worked $3\frac{1}{3}$ days alone and then Nichole and Richard worked together to finish the job, how many days did they work together?

1 is the total amount of the job. Let x be the number of days they work together. Then, Nichole will do $\frac{1}{8}$ of the job in a day and Richard will do $\frac{1}{6}$ of the job in a day.

So, $\frac{3\frac{1}{3}}{8} + \frac{1}{8}x + \frac{1}{6}x = 1$; $\frac{10+3x+4x}{24} = 1$; $7x = 24 - 10 = 14$; $x = 2$

Therefore, they worked together for 2 days.

 # 35 Nichole checked a book out from a library. She read $\frac{1}{3}$ of the book on the first day, $\frac{1}{4}$ of the book on the second day, and 39 pages on the third day. She now has to read $\frac{1}{5}$ of the book to finish. How many pages does the book have?

Let x be the number of pages. Then,

$\frac{1}{3}x + \frac{1}{4}x + 39 = \frac{4}{5}x$; $\frac{7}{12}x + 39 = \frac{4}{5}x$; $\frac{13}{60}x = 39$; $x = 180$

Therefore, the book has 180 pages.

 # 36 Richard finishes a job alone in 5 hours. If Nichole helps him, they can finish the job together in 1 hour 40 minutes. How many hours would it take Nichole to work alone to finish the job?

1 is the total amount of the job to finish. Let x be the time in hours Nichole needs to finish the job alone. Since 1 hour 40 minutes is $1\frac{40}{60} = 1\frac{2}{3} = \frac{5}{3}$, we have $\frac{1}{5} + \frac{1}{x} = \frac{3}{5}$.

so, $\frac{x+5}{5x} = \frac{3}{5}$; $5x + 25 = 15x$; $10x = 25$; $x = \frac{5}{2} = 2\frac{1}{2}$ Therefore, 2 hours 30 minutes.

 # 37 Nichole goes out to eat at a restaurant. Her total bill is $23, including a 15% tip. How much was the dinner?

$$x + x \cdot \frac{15}{100} = 23 \; ; \; \frac{115}{100}x = 23 \; ; \; x = 20$$

Therefore, it was $20.

 # 38 A movie ticket price for children is $3 less than the adult ticket price. Nichole paid $36 for 2 adults and 3 children. What is the price of an adult ticket?

$$2x + 3(x - 3) = 36 \; ; \; 5x - 9 = 36 \; ; \; x = 9$$

Therefore, the adult ticket's price is $9.

 # 39 Nichole and Richard live in the same home. They drove to a park to meet some friends. They started from their home at the same time. Nichole drove at 40 miles per hour and Richard drove at 50 miles per hour. Nichole arrived at the park 10 minutes late while Richard arrived 5 minutes early for their appointment. Find the distance from Richard and Nichole's home to the park.

Let x be the distance from Richard and Nichole's home to the park. Then,

$$40t + 40 \cdot \frac{1}{6} = 50t - 50 \cdot \frac{1}{12} \; ; \; 10t = \frac{1}{12}(40 \cdot 2 + 50) \; ; \; t = \frac{130}{12} \cdot \frac{1}{10} = \frac{13}{12}$$

So $\quad x = 40 \left(t + \frac{10}{60}\right) = 40 \left(\frac{13}{12} + \frac{1}{6}\right) = 40 \cdot \frac{15}{12} = 40 \cdot \frac{5}{4} = 50 \quad$ or

$$x = 50 \left(t - \frac{5}{60}\right) = 50 \left(\frac{13}{12} - \frac{1}{12}\right) = 50$$

Therefore, the distance is 50 miles.

 # 40 Find the value for each of the following

(1) $|4| = 4$

(2) $|-5| = -(-5) = 5$

(3) $-|-3| = -(3) = -3$

(4) $|2 - 6| = |-4| = -(-4) = 4$

(5) $|-7 - 5| = |-12| = -(-12) = 12$

(6) $|5| + |-5| = 5 + -(-5) = 5 + 5 = 10$

(7) $|3| - |8| = 3 - 8 = -5$

(8) $|-9| + (-9) = -(-9) - 9 = 9 - 9 = 0$

 # 41 Solve the following equations

(1) $|x - 3| = 5x + 2$

If $x - 3 \geq 0$ ($x \geq 3$) $\Rightarrow$ $|x - 3| = x - 3 = 5x + 2$ $\Rightarrow$ $4x = -5$ $\Rightarrow$ $x = -\frac{5}{4}$

Since $x \geq 3$, $x = -\frac{5}{4}$ is not a solution.

If $x - 3 < 0$ ($x < 3$) $\Rightarrow$ $|x - 3| = -(x - 3) = 5x + 2$ $\Rightarrow$ $6x = 1$ $\Rightarrow$ $x = \frac{1}{6}$

Since $x < 3$, $x = \frac{1}{6}$ is a solution.

Therefore, the solution of $|x - 3| = 5x + 2$ is $x = \frac{1}{6}$

(2) $|x - 4| + |x + 2| = 10$

If $x - 4 = 0$ $\Rightarrow$ $x = 4$

If $x + 2 = 0$ $\Rightarrow$ $x = -2$

$\therefore$ Consider $x < -2$, $-2 \leq x < 4$, and $x \geq 4$

If $x < -2$ $\Rightarrow$ $|x - 4| + |x + 2| = -(x - 4) - (x + 2) = 10$

$\Rightarrow$ $-2x = 8$ $\Rightarrow$ $x = -4$; a solution

If $-2 \leq x < 4$ $\Rightarrow$ $|x - 4| + |x + 2| = -(x - 4) + (x + 2) = 10$

$\Rightarrow$ $0 \cdot x = 4$; undefined

If $x \geq 4$ $\Rightarrow$ $|x - 4| + |x + 2| = (x - 4) + (x + 2) = 10$

$\Rightarrow$ $2x = 12$ $\Rightarrow$ $x = 6$; a solution

Therefore, the solutions are $x = -4$ and $x = 6$.

(3) $|x - 2| - |5 - x| = 0$

$|x - 2| - |5 - x| = 0$ $\Rightarrow$ $|x - 2| = |5 - x|$ $\Rightarrow$ $x - 2 = \pm(5 - x)$

If $x - 2 = 5 - x$ $\Rightarrow$ $2x = 7$ $\Rightarrow$ $x = \frac{7}{2}$

If $x - 2 = -(5 - x)$ $\Rightarrow$ $0 \cdot x = -3$ undefined

Therefore, the solution is $x = \frac{7}{2}$.

Solutions for Chapter 2

#1. Express each statement as an inequality.

(1) a is less than -3 ; $a < -3$

(2) a is greater than or equal to 2 ; $a \geq 2$

(3) a is greater than -1 and less than or equal to 1 ; $-1 < a \leq 1$

(4) 3 more than twice a is greater than half of a ; $2a + 3 > \frac{1}{2}a$

(5) 4 less than three time a is greater than or equal to a plus 2 ; $3a - 4 \geq a + 2$

(6) a is not greater than 0 ; $a \ngtr 0 \therefore a \leq 0$

#2. Solve the following inequalities

(1) $x - 5 > 6$; $x > 6 + 5$; $x > 11$

(2) $x + 4 > 0$; $x > 0 - 4$; $x > -4$

(3) $6x > 3$; $x > \frac{1}{2}$

(4) $2x + 3 > 7$; $2x + 3 - 3 > 7 - 3$; $2x > 4$; $x > 2$

(5) $3x - 4 > x + 3$; $3x - x > 3 + 4$; $2x > 7$; $x > \frac{7}{2}$

(6) $x + 5 > 3x$; $-2x > -5$; $x < \frac{5}{2}$

(7) $-2x - 5 \leq 7$; $-2x \leq 12$; $x \geq -6$

(8) $-\frac{1}{3}x - 1 \leq 8$; $-\frac{1}{3}x \leq 9$; $x \geq -27$

(9) $-2x > 4$; $x < -2$

(10) $-3x + 4 < -2x$; $-3x + 2x < -4$; $-x < -4$; $x > 4$

(11) $2x > 2(x + 3)$; $0 \cdot x > 6$; false $\therefore$ no solution

(12) $5x - (7x - 6) \geq 3$; $-2x + 6 \geq 3$; $-2x \geq -3$; $x \leq \frac{3}{2}$

(13) $3x - (8x + 5) \leq 2$; $-5x - 5 \leq 2$; $-5x \leq 7$; $x \geq -\frac{7}{5}$

(14) $2.5x - 1.5 > 3.5x + 4.5$; $25x - 15 > 35x + 45$; $-10x > 60$; $x < -6$

(15) $2(x + 1) - \frac{8x+1}{3} < 4$; $6(x + 1) - (8x + 1) < 12$; ; $-2x + 5 < 12$; $-2x < 7$; $x > -\frac{7}{2}$

(16) $\frac{4}{3}x - 4\left(\frac{1}{3}x + 2\right) > -1$; $-8 > -1$; false $\therefore$ no solution

(17) $\frac{5x-3}{4} \geq x - \frac{5x+1}{3}$; $3(5x - 3) \geq 12x - 4(5x + 1)$; $23x \geq 5$; $x \geq \frac{5}{23}$

(18) $0.3 - 0.2x < 0.4x - 0.1$; $3 - 2x < 4x - 1$; $-6x < -4$; $x > \frac{2}{3}$

(19) $3 - 2ax < -3$ for $a < 0$; $2ax > 6$; $ax > 3$; since $a < 0$, $x < \frac{3}{a}$

(20) $-ax - 1 \leq 2$ **for** $a < 0$; $-ax \leq 3$; Since $a < 0$, $-a > 0$. $\therefore x \leq -\dfrac{3}{a}$

(21) $2ax > -a$ **for** $a < 0$; $x < \dfrac{-a}{2a}$; $x < -\dfrac{1}{2}$

(22) $3x < 3x + 4$; $0 \cdot x < 4$; always true $\quad \therefore$ All real numbers are the solution.

#3. Solve the following inequalities. Then draw the solution on a number line

(1) $2x - 4 > 4$; $2x > 8$; $x > 4$

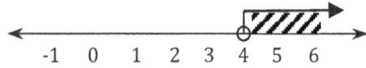

(2) $-3x \leq \dfrac{x-1}{2} - 3$; $-6x \leq (x - 1) - 6$; $-7x \leq -7$; $x \geq 1$

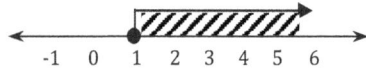

(3) $-\dfrac{x}{4} \geq 2$; $-x \geq 8$; $x \leq -8$

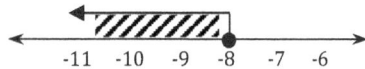

(4) $0.3x - \dfrac{1+x}{2} < -\dfrac{2}{5}$; $3x - 5(1 + x) < -4$; $-2x < 1$; $x > -\dfrac{1}{2}$

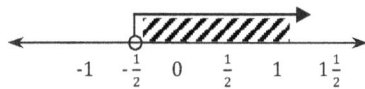

#4. Express the range of x as an inequality.

(1) Three times x minus 5 is greater than five times x plus 2

$$3x - 5 > 5x + 2 \; ; \; -2x > 7 \; ; \; x < -\dfrac{7}{2}$$

(2) Two times the difference of x and 3 is less than or equal to three times the sum of $2x$ and 2

$$2(x - 3) \leq 3(2x + 2) \; ; \; 2x - 6 \leq 6x + 6 \; ; \; -4x \leq 12 \; ; \; x \geq -3$$

#5. Express the range of x for the following expression when $-1 \leq x \leq 1$.

    (1) $2x + 1$; $-2 \leq 2x \leq 2$ $\therefore$ $-1 \leq 2x + 1 \leq 3$

    (2) $-3x - 2$; $3 \geq -3x \geq -3$; $3 - 2 \geq -3x - 2 \geq -3 - 2$ $\therefore$ $-5 \leq -3x - 2 \leq 1$

    (3) $\frac{1}{4}x - 3$; $-\frac{1}{4} \leq \frac{1}{4}x \leq \frac{1}{4}$; $-\frac{1}{4} - 3 \leq \frac{1}{4}x - 3 \leq \frac{1}{4} - 3$ $\therefore$ $-3\frac{1}{4} \leq \frac{1}{4}x - 3 \leq -2\frac{3}{4}$

#6. Let $y = \frac{4 - 2x}{3}$.

    (1) Find the range of y when $1 < x < 5$.

        $-2 > -2x > -10$; $4 - 2 > 4 - 2x > 4 - 10$; $\frac{2}{3} > \frac{4 - 2x}{3} > \frac{-6}{3}$ $\therefore$ $-2 < y < \frac{2}{3}$

    (2) Find the range of y when $-3 < x < -1$.

        $6 > -2x > 2$; $4 + 6 > 4 - 2x > 4 + 2$; $\frac{10}{3} > \frac{4 - 2x}{3} > \frac{6}{3}$; $2 < y < \frac{10}{3}$

    (3) Find the range of x when $2 \leq y \leq 4$.

        $2 \leq \frac{4 - 2x}{3} \leq 4$; $6 \leq 4 - 2x \leq 12$; $6 - 4 \leq -2x \leq 12 - 4$; $\frac{2}{-2} \geq \frac{-2x}{-2} \geq \frac{8}{-2}$

        $-1 \geq x \geq -4$; $-4 \leq x \leq -1$

#7. Find the sum of all positive integers which satisfy the inequality

    $2(1 - x) + 6 \geq 3(x - 3) - 5.$

    $2 - 2x + 6 \geq 3x - 9 - 5$; $5x \leq 8 + 14$; $x \leq \frac{22}{5}$; $x = 1, 2, 3, 4$ $\therefore$ $1 + 2 + 3 + 4 = 10$

#8. The sum of three consecutive integers is greater than or equal to 69. Find the three integers

    with the smallest sum.

    $x + (x + 1) + (x + 2) \geq 69$; $3x + 3 \geq 69$; $3x \geq 66$; $x \geq 22$ $\therefore$ 22, 23, 24

#9. How many positive integers satisfy the following inequalities?

    (1) $3\left(\frac{1}{2}x - 1\right) < x + 2$

        $\frac{3}{2}x - 3 < x + 2$; $\frac{1}{2}x < 5$; $x < 10$; So, $x = 1, 2, \cdots, 9$ $\therefore$ 9 integers

(2) $0.3(2-x) \geq 0.1x - 0.2$; $3(2-x) \geq x - 2$; $-3x + 6 \geq x - 2$; $4x \leq 8$; $x \leq 2$

So, $x = 1, 2$ $\therefore$ 2 integers

(3) $\frac{x+3}{2} - \frac{2-x}{3} < 1$; $3(x+3) - 2(2-x) < 6$; $3x + 9 - 4 + 2x < 6$; $5x < 1$; $x < \frac{1}{5}$

$\therefore$ There are no positive integers that satisfy the inequality.

#10. Only 3 positive integers satisfy the inequality $3x - k \leq \frac{5-x}{2}$. Find the range of k.

Note : When 3 positive integers are satisfying , $\begin{cases} ① & x < k \Rightarrow 3 < k \leq 4 \\ ② & x \leq k \Rightarrow 3 \leq k < 4 \end{cases}$

$3x - k \leq \frac{5-x}{2}$ $\Rightarrow$ $2(3x - k) \leq 5 - x$ $\Rightarrow$ $7x \leq 5 + 2k$ $\Rightarrow$ $x \leq \frac{5+2k}{7}$

$\therefore$ $3 \leq \frac{5+2k}{7} < 4$; $21 \leq 5 + 2k < 28$; $16 \leq 2k < 23$; $8 \leq k < \frac{23}{2}$

#11. Find the constant k if

(1) The inequality $\frac{1}{2}x - \frac{k}{3} < -1$ has the solution $x < 2$.

$3x - 2k < -6$; $3x < 2k - 6$; $x < \frac{2k-6}{3}$

$\therefore$ $\frac{2k-6}{3} = 2$; $2k - 6 = 6$; $2k = 12$ $\therefore k = 6$

(2) The inequality $\frac{kx}{4} - \frac{1}{2} > 1$ has the solution $x < -1$.

$kx - 2 > 4$; $kx > 6$.

Since the direction of the symbol for the solution changes, $k < 0$ and $x < \frac{6}{k}$.

$\therefore \frac{6}{k} = -1$ $\therefore$ $k = -6$

(3) The inequality $\frac{2-kx}{5} - 2 \leq \frac{x}{2} + 1$ has the solution $x \leq -4$.

$2(2 - kx) - 20 \leq 5x + 10$; $(-2k - 5)x \leq 10 + 20 - 4$; $(-2k - 5)x \leq 26$

$x \leq \frac{26}{-2k-5}$.   Since $x \leq -4$, $\frac{26}{-2k-5} = -4$.

$\therefore$ $8k + 20 = 26$; $8k = 6$; $k = \frac{3}{4}$

(4) Two inequalities $2(1 - 2x) - 3 \leq x - 5$ and $\frac{3k - 2x}{3} \leq x + 2k$ have the same solution.

$2(1 - 2x) - 3 \leq x - 5 \Rightarrow -5x \leq -4 ; \quad x \geq \frac{4}{5}$

$\frac{3k - 2x}{3} \leq x + 2k \Rightarrow 3k - 2x \leq 3x + 6k ; \quad 5x \geq -3k ; \quad x \geq -\frac{3k}{5}$

$\therefore -\frac{3k}{5} = \frac{4}{5} ; \quad 3k = -4 \quad \therefore k = -\frac{4}{3}$

(5) The inequality $2 - kx < 2x + k$ has no solution.

$(2 + k)x > 2 - k$.

If $0 \cdot x (= 0) >$ positive number, then there is no solution.

So, $2 + k = 0$ and $2 - k > 0$ (positive number)

$\therefore k = -2$ and $k < 2 \quad \therefore k = -2$

(6) The inequality $-2kx + 5 > 6$ has the solution $x > 2$.

$-2kx + 5 > 6 \Rightarrow 2kx < -1$.

Since the solution is $x > 2$, $k < 0$ and $x > \frac{-1}{2k}$

So, $\frac{-1}{2k} = 2 ; \quad 4k = -1 ; \quad k = -\frac{1}{4}$

(7) $1 - 5x \leq 2x - 5k$ has -2 as a minimum value of the solution.

$1 - 5x \leq 2x - 5k \Rightarrow 7x \geq 1 + 5k \Rightarrow x \geq \frac{1 + 5k}{7}$

$\therefore \frac{1 + 5k}{7}$ is the minimum value of x

So, $\frac{1 + 5k}{7} = -2 ; \quad 1 + 5k = -14 ; \quad 5k = -15 ; \quad k = -3$

(8) The inequality $x - \left(3 + \frac{k}{2}\right) > 2x + k$ has no positive solution.

$x - \left(3 + \frac{k}{2}\right) > 2x + k \Rightarrow -x > k + 3 + \frac{k}{2} \Rightarrow x < -\left(\frac{3}{2}k + 3\right)$

$\therefore -\left(\frac{3}{2}k + 3\right) = 1 ; \quad \frac{3}{2}k + 3 = -1 ; \quad \frac{3}{2}k = -4 ; \quad k = -4 \cdot \frac{2}{3} = -\frac{8}{3}$

#12. The inequality $(-a + 2b)x + b - 3a \leq 0$ has the solution $x \leq -1$. Find the solution for the

inequality $(a - b)x + a - 2b > 0$, where $b > 0$.

$(-a + 2b)x + b - 3a \leq 0 \quad \Rightarrow \quad (-a + 2b)x \leq -b + 3a$

Since the solution is $x \leq -1$ (the direction of symbol is not changed),

$-a + 2b > 0$ and $x \leq \frac{-b+3a}{-a+2b}$ 、

So, $\frac{-b+3a}{-a+2b} = -1$; $-a + 2b = b - 3a$; $2a = -b$; $a = -\frac{1}{2}b$

Substitute $a = -\frac{1}{2}b$ into $(a - b)x + a - 2b > 0 \quad \Rightarrow \quad -\frac{3}{2}bx - \frac{5}{2}b > 0$; $-\frac{3}{2}bx > \frac{5}{2}b$

Since $b > 0$, $-\frac{3}{2}x > \frac{5}{2}$; $x < -\frac{5}{3}$

Note that you cannot use $b = -2a$ instead of $a = -\frac{1}{2}b$.

($\because$ Since $-a + 2b > 0$, $2b > a$. Since $b > 0$, $2b > 0$. So, we have $2b > a$.

Since $a > 0$ or $a < 0$, we cannot decide which condition is necessary for a.

Therefore, we have to use $a = -\frac{1}{2}b$, not $b = -2a$.)

$\therefore$ The solution is $x < -\frac{5}{3}$.

#13. Solve the following inequality for x and graph the solution

(1) $|x - 2| \leq 0$

Since $|a| \geq 0$ for any $a > 0$, $x - 2 = 0$ has to be true to satisfy the inequality.

Therefore, the solution is only $x = 2$.

(2) $|3x + 9| > 0$

Since $|a| \geq 0$ for any $a > 0$, $3x + 9 \neq 0$ has to be true to satisfy the inequality. So, $x \neq -3$

Therefore, the solutions are all real numbers except -3.

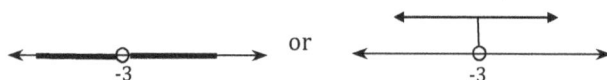

(3) $|x + 4| < 0$

Since $|a| \geq 0$ for any $a > 0$, there is no solution to satisfy the inequality.

(4) $|-2x + 1| + 3 \leq 6$

$$|-2x + 1| \leq 6 - 3 \ ; \ |-2x + 1| \leq 3$$

$\Rightarrow \ -3 \leq -2x + 1 \leq 3$

$\Rightarrow \ -4 \leq -2x \leq 2$

$\Rightarrow \ \dfrac{-4}{-2} \geq x \geq \dfrac{2}{-2}$

$\Rightarrow \ -1 \leq x \leq 2$

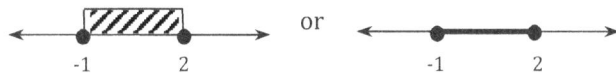

 or

(5) $0 < |2 - 4x| < 8$

(Case 1) If $2 - 4x \geq 0$, then $x \leq \dfrac{1}{2}$

$\Rightarrow \ 0 < 2 - 4x < 8$

$\Rightarrow \ -2 < -4x < 6$

$\Rightarrow \ \dfrac{-2}{-4} > x > \dfrac{6}{-4}$

$\Rightarrow \ -\dfrac{3}{2} < x < \dfrac{1}{2}$

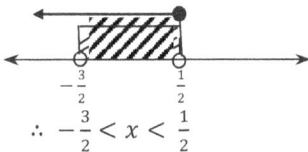

$\therefore \ -\dfrac{3}{2} < x < \dfrac{1}{2}$

(Case 2) If $2 - 4x < 0$, then $x > \dfrac{1}{2}$

$\Rightarrow \ 0 < -(2 - 4x) < 8$

$\Rightarrow \ 0 < -2 + 4x < 8$

$\Rightarrow \ 2 < 4x < 10$

$\Rightarrow \ \dfrac{1}{2} < x < \dfrac{5}{2}$

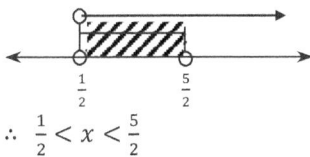

$\therefore \ \dfrac{1}{2} < x < \dfrac{5}{2}$

Therefore, the solutions are $-\frac{3}{2} < x < \frac{1}{2}$ or $\frac{1}{2} < x < \frac{5}{2}$.

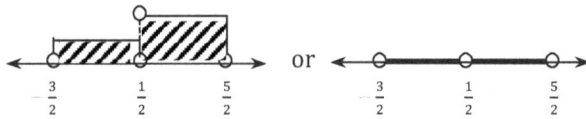

(6) $2 < |x+1| < 3$

(Case 1) If $x+1 \geq 0$, then $x \geq -1$

$\Rightarrow 2 < x+1 < 3$

$\Rightarrow 1 < x < 2$

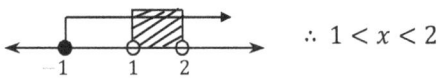 $\therefore 1 < x < 2$

(Case 2) If $x+1 < 0$, then $x < -1$

$\Rightarrow 2 < -(x+1) < 3$

$\Rightarrow 3 < -x < 4$

$\Rightarrow -3 > x > -4$

$\Rightarrow -4 < x < -3$

 $\therefore -4 < x < -3$

Therefore, the solutions are $1 < x < 2$ or $-4 < x < -3$.

#14. At the store there is a bucketful of apples and peaches. An apple is worth 25¢ and a peach is worth 50¢. You want to buy 10 pieces of fruit. What is the maximum number of peaches you can buy with less than $10?

Let x be the number of peaches. Then

$0.25(10 - x) + 0.50x < 10$

$25(10 - x) + 50x < 1000$

$25x < 750$

$x < 30 \qquad \therefore \quad 29$ peaches

#15. Nichole wants to make a salt solution that is at most 10% salt by adding water to 50 ounces of a 15% salt solution. What is the amount of water she can add?

$$\frac{15}{100} \cdot 50 + \frac{0}{100} \cdot x \;\leq\; \frac{10}{100} \cdot (50 + x)$$

$$15 \cdot 50 \leq 10(50 + x)$$

$$10\,x \geq 250$$

$$x \geq 25$$

∴ At least 25 ounces of water.

#16. Richard goes hiking in the mountain. He goes up the trail at a speed of 3 miles per hour and down the same trail at a speed of 4 miles per hour, while hiking for no longer than 2 hours. Find the maximum distance he can hike.

Let x be the distance of the trail.

Then the time for up the hill is $\frac{x}{3}$ and the time for down the hill is $\frac{x}{4}$. So, $\frac{x}{3} + \frac{x}{4} \leq 2$

Thus, $\frac{7x}{12} \leq 2$; $x \leq \frac{24}{7}$; $x \leq 3\frac{3}{7}$

Therefore, the maximum distance is $3\frac{3}{7}$ miles.

#17. Nichole plans to take a 3 miles walk in less than $\frac{1}{2}$ hour. She walks at a speed of 3 miles per hour at the beginning , then runs at a speed of 9 miles per hour for the rest.

How far does she walk?

Let x be the walking distance. Then, $\frac{x}{3}$ is the time for walking and $\frac{3-x}{9}$ is the time for running.

So, $\frac{x}{3} + \frac{3-x}{9} < \frac{1}{2}$; $\frac{3x+3-x}{9} < \frac{1}{2}$; $\frac{2x+3}{9} < \frac{1}{2}$; $2x + 3 < \frac{9}{2}$; $2x < \frac{9}{2} - 3$

$2x < \frac{3}{2}$; $x < \frac{3}{4}$

Therefore, the walking distance is less than $\frac{3}{4}$ mile.

#18. Richard needs to produce a salt solution that is at least 12% salt after mixing 30 ounces of a 5% salt solution with a 15% salt solution. How many more ounces of a 15% salt solution must be needed?

$$\frac{5}{100} \cdot 30 + \frac{15}{100} \cdot x \ > \ \frac{12}{100} \cdot (30 + x)$$

$$150 + 15x > 360 + 12x$$

$$3x > 210$$

$$x > 70$$

∴ More than 70 ounces of a 15% salt solution.

#19. Nichole's last scores on three math tests were $93, 87,$ and 89. When she takes her next test, she wants to have a total average of at least 92 points for all four tests. What score does she need to get on her next test?

Let x be the score Nichole needs to get. Then, $\dfrac{93+87+89+x}{4} \geq 92$

$$93 + 87 + 89 + x \geq 368$$

$$269 + x \geq 368 \ ; \ x \geq 99$$

∴ At least 99 points

#20. Richard is 14 years old and his dad is 48 years old. In how many years will his dad's age become less than twice Richard's age?

Let dad's age be twice Richard's age in x years. Then,

$$48 + x < 2(14 + x)$$

So, $48 + x < 28 + 2x \ ; \ x > 20$

∴ After 21 years

#21. **The lengths of three sides of a triangle are x, $x + 2$, and $x + 3$, where x is a positive integer. Find the smallest length of the triangle.**

Since the longest length of a triangle is less than the sum of other lengths (triangle inequality),

$x + 3 < x + (x + 2)$

$x + 3 < 2x + 2$; $x > 1$ $\therefore$ The smallest length is 2.

#22. **The shaded area is, at most, 66 square inches. Find the smallest integer for x.**

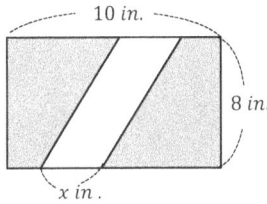

Shaded area $= 80 - 8x$

$\therefore 80 - 8x \not> 66$; $80 - 8x \leq 66$; $8x \geq 14$; $x \geq \frac{14}{8}$

Therefore, the smallest integer is $x = 2$.

#23. **There is a big sale going on at a book store. Nichole buys a book that is 20% off. But her $20 bill is not enough to buy it. What is the price range of the book?**

Let x be the price of the book. Then,

$x - \frac{20}{100}x > 20$; $\frac{80}{100}x > 20$; $x > 20 \cdot \frac{10}{8}$; $x > 25$;

Therefore, the price of the book is more than $25.

#24. **Suppose one boy can complete a task in 4 hours and one girl can complete the same task in 6 hours. A group of 5 boys and girls try to complete the task in 1 hour. Find the minimum number of boys to complete the task.**

Let 1 be the total amount of the task and x be the number of boys.

Thus, $\frac{x}{4} + \frac{5-x}{6} \geq 1$

Thus, $3x + 2(5 - x) \geq 12$; $x \geq 2$

$\therefore$ At least 2 boys are required.

<u>Solutions for Chapter 3</u>

 # 1. Simplify each expression.

(1) $a^2 \cdot a^3 \cdot a^4 = a^{2+3+4} = a^9$

(2) $x^3 \cdot y^2 \cdot x^4 \cdot y \cdot z = x^7 \cdot y^3 \cdot z$

(3) $(2^3 xy^2 z^3)^2 = 2^6 x^2 y^4 z^6$

(4) $(x^3)^2 \cdot (x^4)^3 = x^6 \cdot x^{12} = x^{18}$

(5) $((-x)^2)^3 \cdot ((-x)^3)^2 = x^6 \cdot x^6 = x^{12}$

(6) $(-a^2 b^3)^5 = (-a^2)^5 (b^3)^5 = -a^{10} \cdot b^{15}$

(7) $-3xy^2 (-2x^2 yz^3)^3 = -3xy^2 \cdot (-8x^6 y^3 z^9) = 24x^7 y^5 z^9$

(8) $\left(-\dfrac{x}{y^2}\right)^2 = \dfrac{x^2}{y^4}$

(9) $\dfrac{a^2 a^3}{(-a)^4} = \dfrac{a^5}{a^4} = a^1$

(10) $\left(\dfrac{2}{3}a^2\right)^2 \cdot \left(\dfrac{3}{4}a^3\right)^2 = \dfrac{4}{9}a^4 \cdot \dfrac{9}{16}a^6 = \dfrac{1}{4}a^{10}$

(11) $(-a^2 b)^3 \div (-a)^3 \cdot (ab^2)^2 - \dfrac{-a^6 b^3}{-a^3} \cdot a^2 b^4 = \dfrac{a^8 b^7}{a^3} = a^5 b^7$

(12) $\left(\dfrac{2}{3}\right)^{-3} = \left(\dfrac{3}{2}\right)^3 = \dfrac{27}{8}$

(13) $\left(\dfrac{ab}{a^2 b^3}\right)^2 = \dfrac{a^2 b^2}{a^4 b^6} = \dfrac{1}{a^2 b^4}$

(14) $\dfrac{x^3 x^{-4}}{x^2} = \dfrac{x^{-1}}{x^2} = \dfrac{1}{x^3}$

(15) $\dfrac{a^3 b^{-2}}{a^{-4} b^3} = \dfrac{a^7}{b^5}$

(16) $\dfrac{2^3 + 2^3 + 2^3}{5^2 + 5^2 + 5^2} = \dfrac{3 \cdot 2^3}{3 \cdot 5^2} = \dfrac{8}{25}$

(17) $3^{2a-1} + 3^{2a-1} + 3^{2a-1} = 3 \cdot 3^{2a-1} = 3^{1+2a-1} = 3^{2a} = 9^a$

(18) $\dfrac{3^4 + 3^5 + 3^6 + 3^7}{3 + 3^2 + 3^3 + 3^4} = \dfrac{3^3(3 + 3^2 + 3^3 + 3^4)}{3 + 3^2 + 3^3 + 3^4} = 3^3 = 27$

(19) $\dfrac{4^3 + 4^3 + 4^3 + 4^3}{4^3 \cdot 4^3 \cdot 4^3 \cdot 4^3} = \dfrac{4 \cdot 4^3}{4^{12}} = \dfrac{4^4}{4^{12}} = \dfrac{1}{4^8}$

(20) $-3xy^2 \cdot (-2x^2y)^3 \div (2xy)^2 = \frac{-3xy^2 \cdot -8x^6y^3}{4x^2y^2} = 6x^5y^3$

(21) $\left(-\frac{3}{2}xy^3\right)^3 \div 4x^2y \cdot \left(-\frac{4}{3}x^3y\right)^2 = -\frac{27}{8}x^3y^9 \cdot \frac{1}{4x^2y} \cdot \frac{16}{9}x^6y^2 = -\frac{3}{2}x^7y^{10}$

(22) $3^{-1} \cdot \left(\frac{1}{2}\right)^3 \cdot 3^3 = \frac{1}{3} \cdot \frac{1}{8} \cdot 3^3 = \frac{9}{8}$

(23) $8^{a-1} \cdot 2^{3a+1} \div 4^{3a-1}$

$= (2^3)^{a-1} \cdot 2^{3a+1} \div (2^2)^{3a-1} = 2^{3a-3} \cdot 2^{3a+1} \div 2^{6a-2} = 2^{3a-3+(3a+1)-(6a-2)} = 2^0 = 1$

2. Find all expressions that are true. (4), (6), and (8)

(1) $(a^2)^3 = a^5$; $(a^2)^3 = a^6$ (false)

(2) $(-a)^3 \cdot -a^2 = -a^5$; $-a^3 \cdot -a^2 = a^5$ (false)

(3) $a^3 \div a^3 = a^1$; $a^{3-3} = a^0 = 1$ (false)

(4) $a^4 \div a^3 \cdot a^5 = a^6$; $a^{4-3+5} = a^6$ (true)

(5) $a^2 + a^3 = a^5$; $a^2 + a^3 = a^2(1+a)$ (false)

(6) $a^{-2} \cdot b^{-2} = (ab)^{-2}$; $\frac{1}{a^2} \cdot \frac{1}{b^2} = \frac{1}{(ab)^2} = (ab)^{-2}$ (true)

(7) $\left(\frac{a}{b^2}\right)^3 = \frac{a^3}{b^5}$; $\left(\frac{a}{b^2}\right)^3 = \frac{a^3}{b^6}$ (false)

(8) $(a^2b)^3 \div -2ab = -\frac{1}{2}a^5b^2$; $(a^2b)^3 \div -2ab = \frac{a^6b^3}{-2ab} = -\frac{1}{2}a^5b^2$ (true)

(9) $(a^2)^3 = a^{2^3}$; $(a^2)^3 = a^6$ (false)

(10) $\left(\frac{3}{x}\right)^2 = \frac{1}{x^2}$; $\frac{3^2}{x^2} = \frac{9}{x^2}$ (false)

(11) $(2a^2)^3 = 6a^6$; $(2a^2)^3 = 8a^6$ (false)

3. Find a and b for the following

(1) $32^3 = (2^a)^3 = 2^b$; $(2^5)^3 = (2^a)^3 = 2^b$; $32 = 2^5$; $a = 5, b = 15$

(2) $2^{a+3} = 8^3$; $2^{a+3} = 8^3 = (2^3)^3 = 2^9$; $a+3 = 9$; $a = 6$

(3) $(2^3)^2 \cdot (2^4)^a = 2^{18}$; $2^6 \cdot 2^{4a} = 2^{18}$; $6 + 4a = 18$; $a = 3$

(4) $(3^b)^3 \div 3^5 = 3^{10}$; $3^{3b-5} = 3^{10}$; $3b - 5 = 10$; $b = 5$

(5) $(4^3)^a = 2^{42}$; $2^{6a} = 2^{42}$; $6a = 42$; $a = 7$

(6) $24^4 = 2^a \cdot 3^b$; $(3 \cdot 2^3)^4 = 3^4 \cdot 2^{12}$; $a = 12$, $b = 4$

(7) $16^a = 2^{a+3}$; $(2^4)^a = 2^{4a} = 2^{a+3}$; $4a = a + 3$; $a = 1$

(8) $(2^3)^4 \div 8^3 \cdot (3^3)^2 = 2^a \cdot 3^b$; $\frac{2^{12} \cdot 3^6}{2^9} = 2^3 \cdot 3^6$; $a = 3, b = 6$

(9) $(2)^{3a+1} \div (2)^{2a-3} = 4$; $\frac{2^{3a} \cdot 2}{2^{2a} \cdot 2^{-3}} = 2^a \cdot 2^4 = 2^{a+4} = 2^2$; $a = -2$

(10) $5^a + 5^{a+2} = 3250$; $5^a + 5^{a+2} = 5^a(1 + 5^2) = 3250$; $5^a = \frac{3250}{26} = 125 = 5^3$; $a = 3$

(11) $2^a + 2^{a+2} = 160$; $2^a + 2^{a+2} = 2^a(1 + 4) = 160$; $2^a = \frac{160}{5} = 32 = 2^5$; $a = 5$

(12) $2^{a-2} = 0.5^{2a-1}$; $2^{a-2} = 0.5^{2a-1} = (\frac{1}{2})^{2a-1} = 2^{-2a+1}$; $a - 2 = -2a + 1$; $a = 1$

(13) $(-2x^a)^b = -32x^{15}$; $(-2)^b x^{ab} = -32x^{15} = (-2)^5 x^{15}$; $ab = 15, b = 5$; $a = 3, b = 5$

(14) $2^{a+2} = 2^{a+1} + 8$; $2^{a+2} - 2^{a+1} - 8 = 0$; $2^a \cdot 4 - 2^a \cdot 2 - 8 = 0$; $2^a(4 - 2) = 8$; $2^a = 4$; $a = 2$

(15) $(x^3 y)^2 \cdot (xy^2)^a \div x^2 y^3 = x^b y^{13}$; $\frac{x^6 y^2 \cdot x^a y^{2a}}{x^2 y^3} = x^{4+a} y^{2a-1} = x^b y^{13}$; $4 + a = b, 2a - 1 = 13$

$\therefore a = 7, b = 11$

 # 4. $a = 2^{x+1}, b = 3^{x-1}$. **Express 6^x using a and b.**

$6^x = (2 \cdot 3)^x = 2^x \cdot 3^x$

Since $a = 2^{x+1} = 2^x \cdot 2$, $2^x = \frac{a}{2}$

Since $b = 3^{x-1} = 3^x \cdot 3^{-1} = \frac{3^x}{3}$, $3^x = 3b$ Therefore, $6^x = \frac{a}{2} \cdot 3b = \frac{3}{2} ab$

 # 5. $10^x = 2$, $10^y = 3$. Simplify $6^{\frac{x-y}{x+y}}$.

$6 = 2 \cdot 3 = 10^x \cdot 10^y = 10^{x+y}$

$6^{\frac{x-y}{x+y}} = (10^{x+y})^{\frac{x-y}{x+y}} = 10^{x-y} = 10^x \div 10^y = \frac{2}{3}$

 # 6. For a positive integer n, compute $(-1)^{2n+1} \cdot (-1)^{3n-1} \cdot (-1)^{2n-1} \div (-1)^{3n}$

$(-1)^{2n+1} \cdot (-1)^{3n-1} \cdot (-1)^{2n-1} \div (-1)^{3n}$

$= \frac{(-1)^{2n+1} \cdot (-1)^{3n-1} \cdot (-1)^{2n-1}}{(-1)^{3n}} = \frac{(-1)^{2n+1+3n-1+2n-1}}{(-1)^{3n}} = \frac{(-1)^{7n-1}}{(-1)^{3n}} = (-1)^{4n-1} = \frac{(-1)^{4n}}{(-1)}$

$= \frac{((-1)^4)^n}{-1} = \frac{1^n}{-1} = -1^n = -1$

7. Order the following numbers from least to greatest 2^{32}, 4^{10}, 8^7, $\left(\frac{1}{2}\right)^{-30}$

2^{32}, $4^{10} = (2^2)^{10} = 2^{20}$, $8^7 = (2^3)^7 = 2^{21}$, $\left(\frac{1}{2}\right)^{-30} = (2^{-1})^{-30} = 2^{30}$

Therefore, the correct order is $4^{10}, 8^7, \left(\frac{1}{2}\right)^{-30}, 2^{32}$

8. Find the sum of all possible values of a natural number a which satisfies $a^{2a-1} = a^{3a-4}$.

If $a = 1$, then $a^{2a-1} = a^{3a-4}$ is always true.

If $a \neq 1$, then $2a - 1 = 3a - 4$; $a = 3$

So, the sum is 4.

9. For any positive number n, $2^{n+3}(3^{n+1} + 3^{n+2}) = a6^n$. Find the value of a.

$2^{n+3}(3^{n+1} + 3^{n+2}) = (2^n \cdot 2^3)(3 \cdot 3^n + 9 \cdot 3^n) = 8 \cdot 2^n(12 \cdot 3^n) = 8 \cdot 12 \cdot (2 \cdot 3)^n = 96 \cdot 6^n$

So, $a6^n = 96 \cdot 6^n$ Therefore, $a = 96$

10. For a solid with a length of a^3b^4, width of $3ab^2$, and volume of $15\,a^7b^8$, find the height.

$15\,a^7b^8 = a^3b^4 \cdot 3ab^2 \cdot \text{height}$; $\text{height} = \frac{15\,a^7b^8}{a^3b^4 \cdot 3ab^2} = 5a^3b^2$

11. Find the number of digits in the final value of the following expressions

(1) $2^4 \cdot 3^2 \cdot 5^5$

Note : $2^m \cdot 5^n = 10^l \cdot a$ for positive numbers $m, n,$ and l

For example, $2^5 \cdot 5^3 = 2^3 \cdot 2^2 \cdot 5^3 = (2 \cdot 5)^3 \cdot 2^2 = 10^3 \cdot 4 = 4000$

Since $2 \cdot 5 = 10$, consider $2^4 \cdot 5^5$

Note $4 < 5$; $2^4 \cdot 5^5 = 2^4 \cdot 5^4 \cdot 5^1$

$2^4 \cdot 3^2 \cdot 5^5 = 2^4 \cdot 3^2 \cdot 5^4 \cdot 5^1 = (2 \cdot 5)^4 \cdot 3^2 \cdot 5^1 = 10^4 \cdot 9 \cdot 5 = 45 \cdot 10^4 = 450000$; 6 digits

(2) $5^2 \cdot 3^3 \cdot 20 \cdot 6$

$5^2 \cdot 3^3 \cdot 20 \cdot 6 = 5^2 \cdot 3^3 \cdot 2 \cdot 10 \cdot 2 \cdot 3 = (5 \cdot 2)^2 \cdot 3^4 \cdot 10 = 10^2 \cdot 3^4 \cdot 10 = 3^4 \cdot 10^3 = 81 \cdot 10^3$; 5 digits

(3) $4^8 \cdot 5^{10}$

$(2^2)^8 \cdot 5^{10} = 2^{16} \cdot 5^{10} = 2^{10} \cdot 2^6 \cdot 5^{10} = (2 \cdot 5)^{10} \cdot 2^6 = 64 \cdot 10^{10}$; 12 digits

12. Simplify each polynomial.

(1) $(2a + 3b) - (a - b) = 2a + 3b - a + b = a + 4b$

(2) $(3a^2 - a + 3) - (-5a + 3) = 3a^2 - a + 3 + 5a - 3 = 3a^2 + 4a$

(3) $(2a + 3) - (a^2 - 2a + 5) = 2a + 3 - a^2 + 2a - 5 = -a^2 + 4a - 2$

(4) $4x - \{-2y + 3x - (2x - y) + 3\} - (x - 3y)$

$\quad = 4x + 2y - 3x + 2x - y - 3 - x + 3y = 2x + 4y - 3$

(5) $\frac{1}{3}x - \frac{2}{3}y - (2x + 3y) - \frac{1}{2}x = \frac{1}{3}x - \frac{2}{3}y - 2x - 3y - \frac{1}{2}x = \frac{2x - 12x - 3x}{6} - \left(\frac{2}{3} + \frac{9}{3}\right)y = -\frac{13}{6}x - \frac{11}{3}y$

(6) $\frac{x+3y-1}{2} - \frac{2x-y+2}{3} = \frac{3x+9y-3-4x+2y-4}{6} = \frac{-x+11y-7}{6} = -\frac{1}{6}x + \frac{11}{6}y - \frac{7}{6}$

(7) $2a - [a^2 - \{3b - (2a - b) + a^2\} - 5] = 2a - [a^2 - 3b + 2a - b - a^2 - 5]$

$\quad = 2a - a^2 + 3b - 2a + b + a^2 + 5 = 4b + 5$

13. When an integer a is divided by 5, the remainder is 1. When an integer b is divided by 5, the remainder is 2. Find the remainder when $a + b$ is divided by 5.

Let x and y be the quotients of a and b, respectively.

$a = 5x + 1$, $b = 5y + 2$; $a + b = 5(x + y) + 3$ $\therefore$ The remainder of $a + b$ is 3.

14. a is the coefficient of x^2 and b is the constant of the following polynomials. Find $a - b$.

(1) $-(2x^2 - 4x + 5) + (3x^2 - x + 1) = -2x^2 + 4x - 5 + 3x^2 - x + 1 = x^2 + 3x - 4$

$\quad a = 1, b = -4$ $\therefore a - b = 1 - (-4) = 5$

(2) $\left(\frac{1}{3}x^2 - \frac{1}{2}x + 2\right) - \left(\frac{1}{2}x^2 - \frac{2}{3}x + 5\right) = \left(\frac{1}{3} - \frac{1}{2}\right)x^2 + \left(-\frac{1}{2} + \frac{2}{3}\right)x - 3 = \frac{2-3}{6}x^2 + \frac{-3+4}{6}x - 3$

$\quad = -\frac{1}{6}x^2 + \frac{1}{6}x - 3$; $a = -\frac{1}{6}, b = -3$ $\therefore a - b = -\frac{1}{6} + 3 = \frac{17}{6}$

(3) $2x^2 - \{3x - (3x^2 + 2x)\} - 2x + 5 = 2x^2 - 3x + 3x^2 + 2x - 2x + 5 = 5x^2 - 3x + 5$

$\quad a = 5, b = 5$ $\therefore a - b = 0$

 # 15. You wanted to add the polynomial $-2a^2 + 3a - 4$ to a polynomial A, but you accidentally subtracted the polynomial from A and got $-3a^2 + 5$. Compute the right answer.

$$A - (-2a^2 + 3a - 4) = -3a^2 + 5 \quad \therefore A = -2a^2 + 3a - 4 - 3a^2 + 5 = -5a^2 + 3a + 1$$

$$\therefore \quad (-5a^2 + 3a + 1) + (-2a^2 + 3a - 4) = -7a^2 + 6a - 3$$

 # 16. If you subtract the polynomial $2a^2 - a + 3$ from two times a polynomial A, then you get $-2a^2 - a - 2$. If you add two times the polynomial $2a^2 - a + 3$ to a polynomial A, then you get $4a^2 + a + 2$. Find the value of a satisfying the two conditions.

$$2A - (2a^2 - a + 3) = -2a^2 - a - 2 \; ; \; 2A = -2a + 1 \; ; \; A = -a + \frac{1}{2}$$

$$A + 2(2a^2 - a + 3) = 4a^2 + a + 2 \; ; \; A = 2a - 6 + a + 2 = 3a - 4$$

Therefore, $-a + \frac{1}{2} = 3a - 4 \; ; \; 4a = \frac{1}{2} + 4 = \frac{9}{2} \quad \therefore \; a = \frac{9}{8}$

 # 17. Find the perimeter of the following shapes.

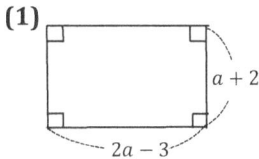

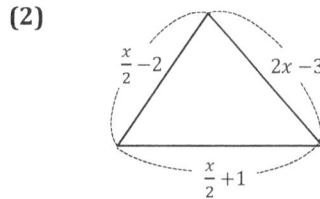

(1) $(2a - 3) + (2a - 3) + (a + 2) + (a + 2) = 6a - 2$

(2) $\left(\frac{x}{2} + 1\right) + \left(\frac{x}{2} - 2\right) + (2x - 3) = 3x - 4$

 # 18. Simplify each polynomial.

(1) $-2x(3x + 4y - 2) = -6x^2 - 8xy + 4x$

(2) $x(x^2 - xy + y^2) - y(-x^2 + xy + y^2) = x^3 - x^2y + xy^2 + yx^2 - xy^2 - y^3 = x^3 - y^3$

(3) $(3a + 2b - 4ab) \cdot -\frac{1}{2}a = -\frac{3}{2}a^2 - ab + 2a^2b$

(4) $(a^2b - ab^2) \div (-ab) = -a + b$

(5) $(3a^2b - 2ab^2) \div \left(-\frac{2}{3}ab\right) = (3a^2b - 2ab^2) \cdot -\frac{3}{2ab} = -\frac{9}{2}a + 3b$

(6) $2a(a - 1) - (a^2 - 1) - 2a(-a + 1) = 3a^2 - 4a + 1$

(7) $-\frac{5}{6}x^2y \cdot \left(\frac{3}{5}xy^2 - 3xy\right) = -\frac{1}{2}x^3y^3 + \frac{5}{2}x^3y^2$

(8) $\left(\frac{2}{3}xy^2 - 2x^2y^2\right) \cdot \left(-\frac{3}{2}xy\right) + \left(\frac{4}{3}x^2y - xy^2\right) \div \left(-\frac{3}{2}xy\right)$

$= \frac{2}{3}xy^2 \cdot \left(-\frac{3}{2}xy\right) + 3x^3y^3 + \dfrac{\frac{4}{3}x^2y - xy^2}{-\frac{3}{2}xy} = -x^2y^3 + 3x^3y^3 + \left(-\frac{8}{9}x + \frac{2}{3}y\right)$

$= -x^2y^3 + 3x^3y^3 - \frac{8}{9}x + \frac{2}{3}y$

(9) $\dfrac{4x^2 - 2xy}{2xy} - \dfrac{6xy^2 - 9y^2}{3xy} = \dfrac{2x}{y} - 1 - 2y + \dfrac{3y}{x}$

(10) $(6x^2y + 3x^2) \div 3x - (3xy - 9y^2) \div 3y = 2xy + x - x + 3y = 2xy + 3y$

(11) $\left(\frac{1}{8}ab - \frac{1}{2}a\right) \cdot 4b - \left(\frac{3}{4}a^2b^2 + a^2b\right) \div 3a$

$= \frac{1}{2}ab^2 - 2ab - \frac{3}{4}a^2b^2 \cdot \frac{1}{3a} - a^2b \cdot \frac{1}{3a} = \frac{1}{2}ab^2 - 2ab - \frac{1}{4}ab^2 - \frac{1}{3}ab = \frac{1}{4}ab^2 - \frac{7}{3}ab$

(12) $\left\{\frac{1}{2}x^2 - \frac{2}{3}(x - 3)\right\} + 3\left\{\frac{1}{2}(x - 2) - \frac{1}{3}(x^2 + 3) + 2\right\}$

$= \frac{1}{2}x^2 - \frac{2}{3}x + 2 + \frac{3}{2}x - 3 - x^2 - 3 + 6 = -\frac{1}{2}x^2 + \left(-\frac{2}{3} + \frac{3}{2}\right)x + 2 = -\frac{1}{2}x^2 + \frac{5}{6}x + 2$

 # 19. $(2x^2y^3)^a \div 4xy \cdot \frac{1}{2}x^2y = bx^3y^3$. **Find the value of $a + b$, where a and b are constants.**

$(2x^2y^3)^a \div 4xy \cdot \frac{1}{2}x^2y = \dfrac{2^a x^{2a} y^{3a} \cdot x^2 y}{4xy \cdot 2} = \dfrac{2^{a-1} x^{2a+2-1} y^{3a+1-1}}{4}$

$= \dfrac{2^{a-1} x^{2a+2-1} y^{3a+1-1}}{2^2} = 2^{a-1-2} x^{2a+1} y^{3a} = bx^3y^3$

$3a = 3 \quad \therefore \quad a = 1, \quad 2^{a-1-2} = 2^{1-1-2} = 2^{-2} = \frac{1}{4} \quad \therefore \quad b = \frac{1}{4}$

Therefore, $a + b = 1 + \frac{1}{4} = \frac{5}{4}$

 # 20. Find the value of $a + b + c$ for the following, where $a, b,$ and c are constants:

(1) $\left(\frac{4}{3}x^2y - 3xy^2 + 2xy\right) \div \frac{1}{2}xy = ax + by + c$

$\left(\frac{4}{3}x^2y - 3xy^2 + 2xy\right) \div \frac{1}{2}xy = \dfrac{2 \cdot \frac{4}{3}x^2y}{xy} - \dfrac{6xy^2}{xy} + 4 = \frac{8}{3}x - 6y + 4$

$\therefore a + b + c = \frac{8}{3} - 6 + 4 = \dfrac{8 - 18 + 12}{3} = \frac{2}{3}$

(2) $\frac{1}{2}(x^2 - 3x + 1) - 2x(x - 1) + 3(4x^2 - 3x - 2) = ax^2 + bx + c$

$\left(\frac{1}{2} - 2 + 12\right)x^2 + \left(-\frac{3}{2} + 2 - 9\right)x + \frac{1}{2} - 6 = 10\frac{1}{2}x^2 - 8\frac{1}{2}x - 5\frac{1}{2}$

$\therefore a + b + c = 10\frac{1}{2} - 8\frac{1}{2} - 5\frac{1}{2} = -3\frac{1}{2}$

 # 21. Find the polynomial for each expression.

(1)   If a polynomial is multiplied by $2ab$, the result is $\frac{1}{2}a^2b + ab^2 - \frac{1}{3}ab$.

Let A be the polynomial. Then,

$$A \cdot 2ab = \frac{1}{2}a^2b + ab^2 - \frac{1}{3}ab \quad \therefore A = \frac{\frac{1}{2}a^2b + ab^2 - \frac{1}{3}ab}{2ab} = \frac{1}{4}a + \frac{1}{2}b - \frac{1}{6}$$

(2)   If a polynomial is divided by $3a - 2b$, the quotient is $\frac{1}{4}ab$ and there is no remainder.

$$\frac{A}{3a-2b} = \frac{1}{4}ab \quad \therefore A = (3a - 2b) \cdot \frac{1}{4}ab = \frac{3}{4}a^2b - \frac{1}{2}ab^2$$

 # 22. Find the area of the shaded part in the rectangle.
The rectangle has a length of $4a$ and width of $2b$.

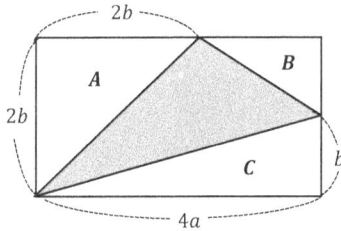

Area of $A = \frac{1}{2} \cdot 2b \cdot 2b = 2b^2$, Area of $B = \frac{(4a-2b)\cdot b}{2} = 2ab - b^2$, Area of $C = \frac{1}{2} \cdot 4a \cdot b = 2ab$

Since the area of the rectangle is $4a \cdot 2b = 8ab$,

the shaded area is $8ab - (2b^2 + 2ab - b^2 + 2ab) = 8ab - b^2 - 4ab = 4ab - b^2$

#23. Expand and simplify each polynomial.

(1) $(2x - 5)(x + 3) = 2x^2 + x - 15$

(2) $(2x - 1)(3x^2 - x - 2) = 6x^3 - 2x^2 - 4x - 3x^2 + x + 2 = 6x^3 - 5x^2 - 3x + 2$

(3) $\left(x + \frac{1}{3}\right)\left(x - \frac{1}{2}\right) = x^2 + \left(\frac{1}{3} - \frac{1}{2}\right)x - \frac{1}{6} = x^2 - \frac{1}{6}x - \frac{1}{6}$

(4) $(3 - 2a)(3 + 2a) = 9 - 4a^2$

(5) $(3a - 2b)(3a + 2b) = 9a^2 - 4b^2$

(6) $(2x + 3)(2x + 3) = 4x^2 + 12x + 9$

(7) $(-2x + 3)(-2x - 3) = 4x^2 - 9$

(8) $(a^3 + b^3)(a^3 - b^3) = a^6 - b^6$

(9) $\left(-4x - \frac{1}{2}\right)^2 = 16x^2 + 4x + \frac{1}{4}$

(10) $(x + y - 2)^2$

Letting $x + y = A$,

$(x + y - 2)^2 = (A - 2)^2 = A^2 - 4A + 4$

$= (x + y)^2 - 4(x + y) + 4 = x^2 + 2xy + y^2 - 4x - 4y + 4$

(11) $102 \times 98 = (100 + 2)(100 - 2) = 100^2 - 2^2 = 10000 - 4 = 9996$

(12) $92 \times 93 = (90 + 2)(90 + 3) = 90^2 + 5 \cdot 90 + 6 = 8100 + 450 + 6 = 8556$

(13) $99^2 = (100 - 1)^2 = 100^2 - 2 \cdot 100 + 1 = 10000 - 200 + 1 = 9801$

(14) $(x + 2y + 3z)(x + 2y - 3z)$

Letting $x + 2y = A$,

$(A + 3z)(A - 3z) = A^2 - 9z^2 = x^2 + 4xy + 4y^2 - 9z^2$

(15) $(2x + y - 3)(2x - y + 3)$

Letting $y - 3 = A$,

$(2x + A)(2x - A) = 4x^2 - A^2 = 4x^2 - (y - 3)^2 = 4x^2 - y^2 + 6y - 9$

(16) $111 \times 109 - 107 \times 113$

$= (110 + 1)(110 - 1) - (110 - 3)(110 + 3) = (110^2 - 1) - (110^2 - 9) = -1 + 9 = 8$

(17) $(2a + b)^2 - (2a - b)^2 = (2a + b + 2a - b)(2a + b - 2a + b) = 4a \cdot 2b = 8ab$

(18) $(a - 3)(a + 2)(a - 1)(a + 4) = (a^2 + a - 12)(a^2 + a - 2) = (A - 12)(A - 2)$

$= A^2 - 14A + 24 = a^4 + 2a^3 + a^2 - 14a^2 - 14a + 24 = a^4 + 2a^3 - 13a^2 - 14a + 24$

24. Find the area of the shaded part of each shape.

(1)

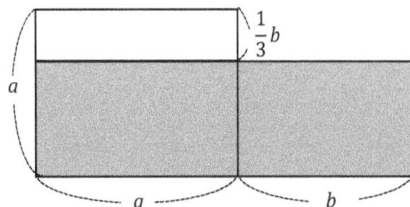

$(a + b)\left(a - \frac{1}{3}b\right) = a^2 + ab - \frac{1}{3}ab - \frac{1}{3}b^2$

$= a^2 + \frac{2}{3}ab - \frac{1}{3}b^2$

(2)

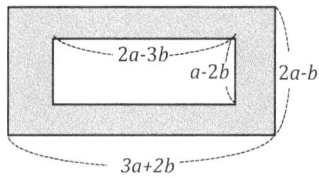

$(3a + 2b)(2a - b) - (2a - 3b)(a - 2b)$

$= 6a^2 + 4ab - 3ab - 2b^2 - (2a^2 - 3ab - 4ab + 6b^2)$

$= 4a^2 + 8ab - 8b^2 = 4(a^2 + 2ab - 2b^2)$

(3)

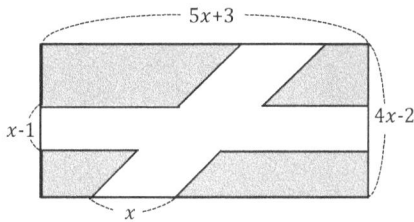

$(5x + 3)(4x - 2) - \{(x - 1)(5x + 3) + x(4x - 2)\} + x(x - 1)$

$= 20x^2 + 2x - 6 - \{5x^2 - 2x - 3 + 4x^2 - 2x\} + x^2 - x$

$= 12x^2 + 5x - 3$

 # 25. Evaluate the polynomial for the variable in each expression.

(1) $3x - 2$ **for** $x = -2$; $3x - 2 = -6 - 2 = -8$

(2) $\frac{2}{3}x + 3$ **for** $x = -1$; $\frac{2}{3}x + 3 = -\frac{2}{3} + 3 = 2\frac{1}{3}$

(3) $-2x^2 - 3x + 1$ **for** $x = -3$; $-2x^2 - 3x + 1 = -2 \cdot 9 + 9 + 1 = -18 + 10 = -8$

(4) $(2x - 2)(-3x + 1)$ **for** $x = 2$; $(2 \cdot 2 - 2)(-3 \cdot 2 + 1) = 2 \cdot -5 = -10$

 # 26. Find the values of the following polynomials

(1) $a^2 + \frac{1}{b^2}$ **when** $a - \frac{1}{b} = 3, \frac{b}{a} = -\frac{1}{3}$; $a^2 + \frac{1}{b^2} = \left(a - \frac{1}{b}\right)^2 + 2a \cdot \frac{1}{b} = 9 + 2 \cdot -3 = 3$

(2) $a^2 + \frac{1}{a^2}$ **when** $a + \frac{1}{a} = 3$; $a^2 + \frac{1}{a^2} = \left(a + \frac{1}{a}\right)^2 - 2a \cdot \frac{1}{a} = 9 - 2 = 7$

(3) ab **when** $a - b = 4, a^2 + b^2 = 8$; $16 = (a - b)^2 = a^2 + b^2 - 2ab = 8 - 2ab$; $ab = -4$

(4) $\left(a - \frac{1}{a}\right)^2$ when $a + \frac{1}{a} = -3$; $\left(a - \frac{1}{a}\right)^2 = \left(a + \frac{1}{a}\right)^2 - 4 = (-3)^2 - 4 = 5$

(5) $\frac{3a+3b}{2a-4b}$ when $\frac{a-b}{a+b} = \frac{2}{3}$

since $\frac{a-b}{a+b} = \frac{2}{3}$, $3(a-b) = 2(a+b)$; $a = 5b$

$\therefore \quad \frac{3a+3b}{2a-4b} = \frac{3\cdot 5b + 3b}{2\cdot 5b - 4b} = \frac{18b}{6b} = 3$

(6) $\frac{b-c}{a} + \frac{c-a}{b} - \frac{a+b}{c}$ when $a + b - c = 0, (abc \neq 0)$

Since $a + b - c = 0$, $b - c = -a$, $c - a = b$, and $a + b = c$.

$\therefore \quad \frac{b-c}{a} + \frac{c-a}{b} - \frac{a+b}{c} = \frac{-a}{a} + \frac{b}{b} - \frac{c}{c} = -1 + 1 - 1 = -1$

(7) $a^4 + b^4$ when $a - b = 1, ab = 2$

$a^4 + b^4 = (a^2 + b^2)^2 - 2a^2b^2 = ((a-b)^2 + 2ab)^2 - 2(ab)^2 = (1 + 4)^2 - 8 = 25 - 8 = 17$

(8) $\frac{(3a-2b)^2}{(2a+3b)^2}$ when $a : b = 3 : 2$

Letting $a = 3k, b = 2k,$

$\frac{(3a-2b)^2}{(2a+3b)^2} = \frac{9a^2 - 12ab + 4b^2}{4a^2 + 12ab + 9b^2} = \frac{81k^2 - 12\cdot 6k^2 + 16k^2}{36k^2 + 12\cdot 6k^2 + 36k^2} = \frac{25\,k^2}{144\,k^2} = \frac{25}{144}$

(9) $\left(\frac{2}{3}a^2 - \frac{3}{2}b^2\right)\left(-\frac{2}{3}a^2 - \frac{3}{2}b^2\right)$ when $a = \frac{1}{2}, b = \frac{1}{3}$

$\left(\frac{2}{3}a^2 - \frac{3}{2}b^2\right)\left(-\frac{2}{3}a^2 - \frac{3}{2}b^2\right) = \left(-\frac{3}{2}b^2\right)^2 - \left(\frac{2}{3}a^2\right)^2 = \frac{9}{4}b^4 - \frac{4}{9}a^4$

$= \frac{9}{4}\left(\frac{1}{3}\right)^4 - \frac{4}{9}\left(\frac{1}{2}\right)^4 = \frac{1}{4}\cdot\frac{1}{9} - \frac{1}{9}\cdot\frac{1}{4} = 0$

(10) $\frac{-3a-6ab+3b}{a-3ab-b}$ when $\frac{1}{a} - \frac{1}{b} = 3$, $ab \neq 0$;

Since $\frac{1}{a} - \frac{1}{b} = 3$, $\frac{b-a}{ab} = 3$; $b - a = 3ab$; $a - b = -3ab$

$\therefore \quad \frac{-3a-6ab+3b}{a-3ab-b} = \frac{-3(a-b)-6ab}{(a-b)-3ab} = \frac{9ab-6ab}{-6ab} = -\frac{1}{2}$

(11) $x^2 + \frac{9}{x^2} - 3$ **when** $x^2 - 4x - 3 = 0$; Since $x^2 - 4x - 3 = 0$, $x - 4 - \frac{3}{x} = 0$

$\left(x - \frac{3}{x}\right)^2 = 4^2$; $x^2 + \frac{9}{x^2} - 6 = 16$; $x^2 + \frac{9}{x^2} = 22$ $\therefore x^2 + \frac{9}{x^2} - 3 = 22 - 3 = 19$

(12) $x^2 + \frac{1}{x^2} - 2x + \frac{2}{x}$ **when** $x^2 + 3x - 1 = 0$

Since $x^2 + 3x - 1 = 0$, $x + 3 - \frac{1}{x} = 0$; $x - \frac{1}{x} = -3$ $\therefore -2x + \frac{2}{x} = 6$

Since $x^2 + \frac{1}{x^2} = \left(x - \frac{1}{x}\right)^2 + 2 = 9 + 2 = 11$, $x^2 + \frac{1}{x^2} - 2x + \frac{2}{x} = 11 + 6 = 17$

(13) $\frac{y}{x} + \frac{x}{y}$ **when** $x - y = 3$, $(x + 2)(y - 2) = -6$

Since $(x + 2)(y - 2) = -6$, $xy + 2(y - x) - 4 = -6$; $xy - 6 - 4 = -6$; $xy = 4$

$\therefore \frac{y}{x} + \frac{x}{y} = \frac{x^2 + y^2}{xy} = \frac{(x-y)^2 + 2xy}{xy} = \frac{9+8}{4} = \frac{17}{4}$

(14) $\frac{3x + 4xy - 3y}{x - y}$ **when** $\frac{1}{x} - \frac{1}{y} = 2$

Since $\frac{1}{x} - \frac{1}{y} = 2$, $\frac{y-x}{xy} = 2$; $y - x = 2xy$; $x - y = -2xy$

$\therefore \frac{3x + 4xy - 3y}{x - y} = \frac{3(x-y) + 4xy}{-2xy} = \frac{-6xy + 4xy}{-2xy} = \frac{-2xy}{-2xy} = 1$

(15) $\frac{x - 2y}{2x + y}$ **when** $\frac{3x+y}{2} = \frac{2x-y}{3}$

Since $\frac{3x+y}{2} = \frac{2x-y}{3}$, $9x + 3y = 4x - 2y$; $5x = -5y$; $x = -y$

$\therefore \frac{x-2y}{2x+y} = \frac{-y-2y}{-2y+y} = \frac{-3y}{-y} = 3$

(16) $\frac{a^2 - b^2}{(a + b)^2}$ **when** $(3x + a)(bx - 1) = (3x - 1)^2$

Since $(3x + a)(bx - 1) = (3x - 1)^2$, $3bx^2 + (ab - 3)x - a = 9x^2 - 6x + 1$

$\therefore a = -1, b = 3$

Therefore, $\frac{a^2 - b^2}{(a+b)^2} = \frac{(a+b)(a-b)}{(a+b)^2} = \frac{a-b}{a+b} = \frac{-1-3}{-1+3} = \frac{-4}{2} = -2$

(17) $\dfrac{1}{xyz}$ **when** $x + \dfrac{1}{y} = 1$, $y + \dfrac{1}{z} = 1$

Since $x + \dfrac{1}{y} = 1$ and $y + \dfrac{1}{z} = 1$, $x = 1 - \dfrac{1}{y} = \dfrac{y-1}{y}$ and $\dfrac{1}{z} = 1 - y$

$\therefore \ \dfrac{1}{xyz} = \dfrac{1}{x} \cdot \dfrac{1}{y} \cdot \dfrac{1}{z} = \dfrac{y}{y-1} \cdot \dfrac{1}{y} \cdot 1 - y = -1$

(18) $(x + 1)(x + 2)(x - 3)(x - 4)$ **when** $x^2 - 2x - 5 = 0$

Since $x^2 - 2x - 5 = 0$, $x^2 - 2x = 5$.

$(x + 1)(x + 2)(x - 3)(x - 4) = (x + 1)(x - 3)(x + 2)(x - 4)$

$= (x^2 - 2x - 3)(x^2 - 2x - 8) = (5 - 3)(5 - 8) = 2 \cdot (-3) = -6$

 # 27. Evaluate each equation for the specified variable.

(1) $2x - 3y + 6 = 0$ **for** x ; $2x = 3y - 6$; $x = \dfrac{3}{2}y - 3$

(2) $x = -2y + 3$ **for** y ; $2y = -x + 3$; $y = -\dfrac{1}{2}x + \dfrac{3}{2}$

(3) $2a = \dfrac{1}{3}(2b - 1)$ **for** b ; $6a = 2b - 1$; $b = 3a + \dfrac{1}{2}$

(1) $c = \dfrac{5}{9}(F \quad 32)$ **for** F , $F - 32 = \dfrac{9}{5}C$, $F = \dfrac{9}{5}C + 32$

(5) $a + b : a - b = 3 : 5$ **for** a ; $5(a + b) = 3(a - b)$; $2a = -8b$; $a = -4b$

(6) $\dfrac{1}{a} + \dfrac{1}{b} = \dfrac{1}{c}, (a \neq 0, b \neq 0, c \neq 0)$ **for** a

$\dfrac{1}{a} = \dfrac{1}{c} - \dfrac{1}{b} = \dfrac{b-c}{cb}$; $a = \dfrac{bc}{b-c}$

(7) $(2a - b)(a + b) = (a + 3b)(2a - b)$ **for** a

$2a^2 + ab - b^2 = 2a^2 + 5ab - 3b^2$; $4ab = 2b^2$; $4a = 2b$; $a = \dfrac{1}{2}b$

(8) $a = -\dfrac{c}{b} + 1$, $b \neq 0, c \neq 0$ **for** b

Since $a = -\dfrac{c}{b} + 1$, $\dfrac{c}{b} = 1 - a$; $b(1 - a) = c$; $b = \dfrac{c}{1-a}$

 # 28. Find the value of $a + b$, where a and b are constants .

(1) $(5 - 1)(5 + 1)(5^2 + 1)(5^4 + 1) = 5^a - b$

$(5 - 1)(5 + 1)(5^2 + 1)(5^4 + 1) = (5^2 - 1)(5^2 + 1)(5^4 + 1)$

$= (5^4 - 1)(5^4 + 1) = (5^8 - 1) = 5^a - b$; $a + b = 8 + 1 = 9$

(2) $8(3^2 + 1)(3^4 + 1)(3^8 + 1) = 3^a + b$

$(3^2 - 1)(3^2 + 1)(3^4 + 1)(3^8 + 1) = (3^4 - 1)(3^4 + 1)(3^8 + 1)$

$= (3^8 - 1)(3^8 + 1) = (3^{16} - 1) = 3^a + b$; $a + b = 16 - 1 = 15$

(3) $(2x + ay)(x - 3y) = 2x^2 - bxy + 9y^2$

$2x^2 + (a - 6)xy - 3ay^2 = 2x^2 - bxy + 9y^2$; $a + b = -3 + 9 = 6$

(4) $(x + y)(x - y) - (3y - 2x)(2x - 3y) = \frac{1}{a}x^2 - 12xy - \frac{1}{b}y^2$

$(x^2 - y^2) + \{(3y - 2x)(3y - 2x)\}$

$= (x^2 - y^2) + \{9y^2 - 12xy + 4x^2\}$

$= 5x^2 - 12xy + 8y^2$; $a = \frac{1}{5}, b = -\frac{1}{8}$; $a + b = \frac{1}{5} - \frac{1}{8} = \frac{3}{40}$

 # 29. Find the sum of the coefficients as well as the constants for each polynomial.

(1) $(x + 2y - 3)(x + 2y - 2)$

Let $A = x + 2y$. Then,

$(A - 3)(A - 2) = A^2 - 5A + 6 = x^2 + 4xy + 4y^2 - 5x - 10y + 6$

∴ $1 + 4 + 4 - 5 - 10 + 6 = 0$

(2) $(2x - 3y - 5)(2x + 3y - 5)$

Let $A = 2x - 5$. Then,

$(A - 3y)(A + 3y) = A^2 - 9y^2 = 4x^2 - 20x + 25 - 9y^2$

∴ $4 - 20 + 25 - 9 = 0$

(3) $(3x + 2y)(x - 2y) - (x + 3y)(x - 2y)$

$(x - 2y)(3x + 2y - x - 3y) = (x - 2y)(2x - y) = 2x^2 - 5xy + 2y^2$

∴ $2 - 5 + 2 = -1$

Solutions for Chapter 4

#1. Solve each system.

(1) $\begin{cases} x + y = 4 \\ 2x - y = 5 \end{cases}$

$$x + y = 4$$
$$+)\ \underline{2x - y = 5}$$
$$3x \quad = 9 \ ; x = 3$$

Substitute $x = 3$; $3 + y = 4$; $y = 1$

$\therefore (x, y) = (3, 1)$

(2) $\begin{cases} 2x + y = 3 \\ 2x + 3y = 4 \end{cases}$

$$2x + y = 3$$
$$-)\ \underline{2x + 3y = 4}$$
$$2y = 1 \ ; \ y = \frac{1}{2}$$

Substitute $y = \frac{1}{2}$; $2x = 3 - y = \frac{5}{2}$; $x = \frac{5}{4}$

$\therefore (x, y) = (\frac{5}{4}, \frac{1}{2})$

(3) $\begin{cases} 2x + y = 7 \\ 3x + 2y = 5 \end{cases}$

$$6x + 3y = 21$$
$$-)\ \underline{6x + 4y = 10}$$
$$y = -11$$

Substitute $y = -11$; $2x = 7 - y = 7 + 11 = 18$; $x = 9$

$\therefore (x, y) = (9, -11)$

(4) $\begin{cases} x - y = 5 \\ 2x + 5y = 3 \end{cases}$

$$2x - 2y = 10$$
$$-)\ \underline{2x + 5y = 3}$$
$$7y = -7 \ ; y = -1$$

Substitute $y = -1$; $x = y + 5 = -1 + 5 = 4$

$\therefore (x, y) = (4, -1)$

(5) $\begin{cases} 3x - y = 0 \\ 5x - 2y = -3 \end{cases}$

Since $3x - y = 0$, $y = 3x$

Substitute $y = 3x$ into $5x - 2y = -3$. Then $5x - 6x = -3$; $x = 3$

$\therefore (x, y) = (3, 9)$

(6) $\begin{cases} 4x - 3y = 6 \\ 6x - 2y = -6 \end{cases}$

$\begin{cases} 4x - 3y = 6 \\ 6x - 2y = -6 \end{cases} \Rightarrow \begin{cases} 4x - 3y = 6 \\ 3x - y = -3 \end{cases} \Rightarrow \begin{cases} 4x - 3y = 6 \\ 9x - 3y = -9 \end{cases}$

$$\begin{array}{r} 4x - 3y = 6 \\ -)\ \underline{9x - 3y = -9} \\ 5x = -15\ ; x = -3 \end{array}$$

Substitute $x = -3$; $y = 3x + 3 = -9 + 3 = -6$

$\therefore (x, y) = (-3, -6)$

#2. Find the value of ab for each system.

(1) $\begin{cases} ax - by = -2 \\ bx + 2y = a \end{cases}$ with solution $(3, 2)$

$\Rightarrow \begin{cases} 3a - 2b = -2 \\ 3b + 4 = a \end{cases}$ $\therefore 3(3b + 4) - 2b = -2$; $9b + 12 - 2b = -2$; $7b = -14$; $b = -2$

So, $a = -6 + 4 = -2$. Therefore, $ab = 4$

(2) $\begin{cases} x + 5y = -3 \\ 2x - by = 5 \end{cases}$ with solution $(a, -1)$

$\Rightarrow \begin{cases} a - 5 = -3 \\ 2a + b = 5 \end{cases} \Rightarrow \begin{cases} a = 2 \\ 2a + b = 5 \end{cases}$ $\therefore b = 5 - 2a = 5 - 4 = 1$

Therefore, $ab = 2$

(3) $\begin{cases} -2x + y = 5 \\ x - 2y = -1 \end{cases}$ with solution (a, b)

$\Rightarrow \begin{cases} -2x + y = 5 \\ 2x - 4y = -2 \end{cases}$

$\Rightarrow -2x + y = 5$

$$\begin{array}{r} -2x + y = 5 \\ +)\ \underline{2x - 4y = -2} \\ -3y = 3\ ;\ y = -1 \end{array}$$

$\therefore x = 2y - 1 = -2 - 1 = -3$

So, $(x, y) = (a, b) = (-3, -1)$ Therefore, $ab = 3$

(4) $\begin{cases} 3x - by = 2 \\ ax + y = -2 \end{cases}$ with solution $(b - 1, 2)$

$\Rightarrow \begin{cases} 3(b-1) - 2b = 2 \\ a(b-1) + 2 = -2 \end{cases} \Rightarrow \begin{cases} b = 5 \\ a(b-1) = -4 \end{cases}$

$\therefore 4a = -4 \; ; a = -1$

Therefore, $ab = -5$

#3. **The system** $\begin{cases} 2x - y = 3 \\ x + 3y = 5 \end{cases}$ **has a solution** $(a + 1, b - 1)$.

Find the value of $(a + b)^2 - (a - b)^2$.

Since $2x - y = 3, y = 2x - 3 \; ; x + 3(2x - 3) = 5 \; ; 7x = 14 \; ; x = 2 \; ; y = 1$

So, $a + 1 = 2, \; b - 1 = 1 \; \therefore \; a = 1, \; b = 2$

$\therefore a + b = 3, \; a - b = -1$

Therefore, $(a + b)^2 - (a - b)^2 = 3^2 - (-1)^2 = 9 - 1 = 8$

#4. **The system** $\begin{cases} 2x + 3y = 5 \\ -x - 2y = -3 \end{cases}$ **has a solution** (a, b).

Find the solution for the system $\begin{cases} (3 - a)x + 2y = -2 \\ 2x + 3y = 2b + 1 \end{cases}$.

$\begin{cases} 2x + 3y = 5 \\ -x - 2y = -3 \end{cases} \Rightarrow \begin{cases} 2x + 3y = 5 \\ -2x - 4y = -6 \end{cases}$

$\Rightarrow \quad 2x + 3y = 5$

$\underline{+) \; -2x - 4y = -6}$

$-y = -1 \; ; \; y = 1$

$\therefore 2x = 5 - 3 = 2 \; ; \; x = 1$ So, $(x, y) = (a, b) = (1, 1)$

$\begin{cases} (3 - a)x + 2y = -2 \\ 2x + 3y = 2b + 1 \end{cases} \Rightarrow \begin{cases} 2x + 2y = -2 \\ 2x + 3y = 3 \end{cases}$

$\Rightarrow \quad 2x + 2y = -2$

$\underline{-) \; 2x + 3y = 3}$

$-y = -5 \; ; \; y = 5$

$\therefore 2x = 3 - 3y = 3 - 15 = -12; \; x = -6$

Therefore, $(x, y) = (-6, 5)$

#5. The solution of the system $\begin{cases} 3x - 2y = -2 \\ (k-1)x + y = -3 \end{cases}$

is the same as the solution of the equation $2x - y = 3$. **Find the constant** k.

Since the solution is the same, rearrange the system.

$$\begin{cases} 3x - 2y = -2 \\ 2x - y = 3 \end{cases} \Rightarrow \begin{cases} 3x - 2y = -2 \\ 4x - 2y = 6 \end{cases}$$

$$\Rightarrow \quad 3x - 2y = -2$$
$$-)\; 4x - 2y = 6$$
$$\overline{\quad\quad -x \quad\quad = -8}\,;\; x = 8$$

Since $2x - y = 3$, $y = 2x - 3 = 16 - 3 = 13$

$\therefore\; (k-1)x + y = -3 \;\Rightarrow\; (k-1)8 + 13 = -3\;;\, 8k - 8 + 13 = -3\;;\, 8k = -8\;;\, k = -1$

Another solution is :

$y = -3 - (k-1)x$ and $y = 2x - 3$ must be the same line. So

$2x - 3 = -3 - (k-1)x\;;\; 2x = (1-k)x$

For these to be the same for all values of x, we must have $k = -1$.

#6. Two systems $\begin{cases} 2x + by = 4 \\ x + 2y = -3 \end{cases}$ **and** $\begin{cases} x - 3y = 2 \\ ax + 2y = -1 \end{cases}$ **have the same solution.**

Find the value of $a + b$.

Let (m, n) be the solution. Then, we have

$$\begin{cases} 2m + bn = 4 \\ m + 2n = -3 \end{cases} \quad \text{and} \quad \begin{cases} m - 3n = 2 \\ am + 2n = -1 \end{cases}$$

Since m and n must satisfy both $m + 2n = -3$ and $m - 3n = 2$, we can solve for m and n by

solving the system $\begin{cases} m + 2n = -3 \\ m - 3n = 2 \end{cases}$

$$\Rightarrow \quad m + 2n = -3$$
$$-)\; m - 3n = 2$$
$$\overline{\quad\quad\quad 5n = -5}\;;\; n = -1$$

Substitute $n = -1$ into $m - 3n = 2$. Then $m = -1$.

So, $(m, n) = (-1, -1)$

Substitute $(m, n) = (-1, -1)$ into $2m + bn = 4$. Then $-2 - b = 4$; $b = -6$

Substitute $(m, n) = (-1, -1)$ into $am + 2n = -1$. Then $-a - 2 = -1$; $a = -1$

$\therefore a + b = -1 - 6 = -7$

#7. Solve each system.

(1) $\begin{cases} x = 2y + 1 \\ x - y = 3 \end{cases}$

$\Rightarrow \begin{cases} x - 2y = 1 \\ x - y = 3 \end{cases}$

$\Rightarrow \quad x - 2y = 1$

$\underline{-)\ x - y = 3}$

$\qquad\qquad -y = -2 \ ; \ y = 2 \ ; \ x = 5 \quad \therefore (x, y) = (5, 2)$

Another solution is :

$x - y = 3 \quad \Rightarrow \quad 2y + 1 - y = 3 \quad \Rightarrow \quad y + 1 = 3 \quad \Rightarrow \quad y = 2$

So $\ x = y + 3 = 2 + 3 = 5 \quad \therefore (x, y) = (5, 2)$

(2) $\begin{cases} y = x - 3 \\ 2x - y = 2 \end{cases}$

$\Rightarrow \begin{cases} -x + y = -3 \\ 2x - y = 2 \end{cases}$

$\Rightarrow \quad -x + y = -3$

$\underline{+)\ 2x - y = 2}$

$\qquad\qquad x \qquad = -1 \ ; \ y = -4$

$\therefore (x, y) = (-1, -4)$

Another solution is :

$2x - y = 2 \quad \Rightarrow \quad 2x - (x - 3) = 2 \Rightarrow \ x + 3 = 2 \quad \Rightarrow \ x = -1$

So $\ y = x - 3 = -1 - 3 = -4 \quad \therefore (x, y) = (-1, -4)$

(3) $\begin{cases} \frac{1}{2}x + \frac{1}{3}y = 2 \\ \frac{2}{3}x - \frac{3}{4}y = -\frac{11}{12} \end{cases}$

$\Rightarrow \begin{cases} 3x + 2y = 12 \\ 8x - 9y = -11 \end{cases} \Rightarrow \begin{cases} 27x + 18y = 108 \\ 16x - 18y = -22 \end{cases}$

$\Rightarrow \quad 27x + 18y = 108$

$\underline{+)\ 16x - 18y = -22}$

$\qquad 43x \qquad\quad = 86 \ ; \ x = 2$

So, $\ \frac{1}{2} \cdot 2 + \frac{1}{3}y = 2 \ ; \ \frac{1}{3}y = 1 \ ; \ y = 3$

$\therefore (x, y) = (2, 3)$

(4)
$$\begin{cases} \dfrac{2}{x} + \dfrac{3}{y} = 4 \\ \dfrac{1}{2x} - \dfrac{1}{y} = 2 \end{cases}$$

Let $A = \dfrac{1}{x}$ and $B = \dfrac{1}{y}$.

$$\Rightarrow \begin{cases} 2A + 3B = 4 \\ \dfrac{1}{2}A - B = 2 \end{cases} \Rightarrow \begin{cases} 2A + 3B = 4 \\ 2A - 4B = 8 \end{cases}$$

$$\Rightarrow \quad 2A + 3B = 4$$
$$-)\ \underline{2A - 4B = 8}$$
$$\qquad 7B = -4\ ;\ B = -\dfrac{4}{7}\ ;\ y = -\dfrac{7}{4}$$

So, $\dfrac{1}{2}A = 2 + B = 2 - \dfrac{4}{7} = \dfrac{10}{7}\ ;\ A = \dfrac{20}{7}\ ;\ x = \dfrac{7}{20}$

$\therefore (x, y) = \left(\dfrac{7}{20}, -\dfrac{7}{4} \right)$

(5)
$$\begin{cases} 0.2x - 0.5y = 0.25 \\ -0.3x + 0.4y = -0.2 \end{cases}$$

$$\Rightarrow \begin{cases} 20x - 50y = 25 \\ -3x + 4y = -2 \end{cases} \Rightarrow \begin{cases} 4x - 10y = 5 \\ -3x + 4y = -2 \end{cases} \Rightarrow \begin{cases} 12x - 30y = 15 \\ -12x + 16y = -8 \end{cases}$$

$$\Rightarrow \quad 12x - 30y = 15$$
$$+)\ \underline{-12x + 16y = -8}$$
$$\qquad -14y = 7\ ;\ y = -\dfrac{1}{2}$$

So, $3x = 4y + 2 = 4\left(-\dfrac{1}{2}\right) + 2 = -2 + 2 = 0$

$\therefore (x, y) = \left(0, -\dfrac{1}{2}\right)$

(6)
$$\begin{cases} 3x - y = -5 \\ 6x - 2y = 3 \end{cases}$$

$$\Rightarrow \begin{cases} 6x - 2y = -10 \\ 6x - 2y = 3 \end{cases}$$

$$\Rightarrow \quad 6x - 2y = -10$$
$$-)\ \underline{6x - 2y = 3}$$
$$\qquad 0 = -13\ ;\ \text{not true} \qquad \therefore \text{no solution}$$

(7) $\begin{cases} x = \frac{1}{2}y + 1 \\ 2x - 4y = -4 \end{cases}$

Since $x = \frac{1}{2}y + 1$, $2\left(\frac{1}{2}y + 1\right) - 4y = -4$; $y + 2 - 4y = -4$; $-3y = -6$; $y = 2$

$\therefore x = 2$ So, $(x, y) = (2, 2)$

(8) $\begin{cases} -\frac{2}{3}x + \frac{1}{4}y = -2 \\ 2x - \frac{3}{4}y = 6 \end{cases}$

$\Rightarrow \begin{cases} 2x - \frac{3}{4}y = 6 \\ 2x - \frac{3}{4}y = 6 \end{cases}$; always true

$\therefore$ unlimited number of solutions

(9) $\begin{cases} \frac{2}{x} + \frac{2}{y} = 1 \\ \frac{1}{x} - \frac{1}{y} = -1 \end{cases}$

Let $A = \frac{1}{x}$ and $B = \frac{1}{y}$.

$\Rightarrow \begin{cases} 2A + 2B = 1 \\ A - B = -1 \end{cases} \Rightarrow \begin{cases} 2A + 2B = 1 \\ 2A - 2B = -2 \end{cases}$

$\Rightarrow \qquad 2A + 2B = 1$

$\qquad +)\ \underline{2A - 2B = -2}$

$\qquad\qquad 4A \qquad = -1$; $A = -\frac{1}{4}$; $x = -4$

$\therefore B = A + 1 = \frac{3}{4}$; $y = \frac{4}{3}$ So, $(x, y) = \left(-4, \frac{4}{3}\right)$

(10) $\frac{x+1}{2} + \frac{y-1}{3} = \frac{2x-1}{3} + \frac{y+2}{4} = 2x + \frac{y}{2}$

$\Rightarrow \begin{cases} \frac{x+1}{2} + \frac{y-1}{3} = 2x + \frac{y}{2} \\ \frac{2x-1}{3} + \frac{y+2}{4} = 2x + \frac{y}{2} \end{cases} \Rightarrow \begin{cases} 3(x+1) + 2(y-1) = 12x + 3y \\ 4(2x-1) + 3(y+2) = 24x + 6y \end{cases} \Rightarrow \begin{cases} 9x + y = 1 \\ 16x + 3y = 2 \end{cases}$

Since $9x + y = 1$, $y = 1 - 9x$ $\therefore 16x + 3(1 - 9x) = 2$; $-11x = -1$; $x = \frac{1}{11}$

So, $y = 1 - 9 \cdot \frac{1}{11} = \frac{2}{11}$

Therefore, $(x, y) = \left(\frac{1}{11}, \frac{2}{11}\right)$

(11) $\begin{cases} 2x - 3y = -4 \\ -3y + z = 2 \\ z + 2x = -6 \end{cases}$

$$2x - 3y = -4$$
$$-3y + z = 2$$
$$+)\ \underline{z + 2x = -6}$$

$$2(2x - 3y + z) = -8\ ;\ 2x - 3y + z = -4$$

Since $2x - 3y = -4$, $z = 0$

Since $z + 2x = -6$, $x = -3$

Since $-3y + z = 2$, $y = -\frac{2}{3}$ $\quad \therefore (x, y, z) = \left(-3, -\frac{2}{3}, 0\right)$

(12) $\begin{cases} x : y + 1 = 2 : 3 \\ 3 : y - 1 = 4 : x - 1 \end{cases}$

$$\Rightarrow \begin{cases} 3x = 2y + 2 \\ 3x - 3 = 4y - 4 \end{cases} \Rightarrow \begin{cases} 3x - 2y = 2 \\ 3x - 4y = -1 \end{cases}$$

$$\Rightarrow \qquad 3x - 2y = 2$$
$$-)\ \underline{3x - 4y = -1}$$

$$2y = 3\ ;\ y = \frac{3}{2} \quad \therefore 3x = 2\left(\frac{3}{2}\right) + 2 = 5\ ;\ x = \frac{5}{3} \quad \text{So, } (x, y) = \left(\frac{5}{3}, \frac{3}{2}\right).$$

#8. The system $\begin{cases} \frac{a+1}{2}x - \frac{3}{4}y = -2 \\ 5x + \frac{b-1}{2}y = 4 \end{cases}$ has an unlimited number of solutions.

Find the value of $a + b$.

$$\frac{\frac{a+1}{2}}{5} = \frac{-\frac{3}{4}}{\frac{b-1}{2}} = \frac{-2}{4} = \frac{-1}{2}$$

So, $2 \cdot \frac{a+1}{2} = -5\ ;\ a + 1 = -5\ ;\ a = -6$

$$-\frac{3}{4} \cdot 2 = -\frac{b-1}{2}\ ;\ -\frac{3}{2} = -\frac{b-1}{2}\ ;\ 3 = b - 1\ ;\ b = 4$$

$$\therefore a + b = -6 + 4 = -2$$

#9. The system $\begin{cases} a(x-y) + \frac{y}{2} = -1 \\ -\frac{x}{2} - \frac{1}{a}y = 3 \end{cases}$ **has no solution. Find the value of a.**

$$\begin{cases} ax + \left(-a + \frac{1}{2}\right)y = -1 \\ -\frac{1}{2}x - \frac{1}{a}y = 3 \end{cases}$$

$$\frac{a}{-\frac{1}{2}} = \frac{-a+\frac{1}{2}}{-\frac{1}{a}} \neq \frac{-1}{3}$$

$$a \cdot -\frac{1}{a} = -\frac{1}{2}\left(-a + \frac{1}{2}\right) \;;\; -1 = -\frac{1}{2}\left(-a + \frac{1}{2}\right) \;;\; -a + \frac{1}{2} = 2 \;;\; a = -\frac{3}{2}$$

#10. The system $\begin{cases} 2kx - (3x + y) = 2y \\ -(k-1)x + 2y = kx \end{cases}$ **has a solution other than $(0, 0)$.**

Find the value of the constant k.

$$\begin{cases} (2k-3)x - 3y = 0 \\ (-k + 1 - k)x + 2y = 0 \end{cases} \;;\; \frac{2k-3}{-k+1-k} = -\frac{3}{2} \;;\; 4k - 6 = 6k - 3 \;;\; 2k = -3 \;;\; k = -\frac{3}{2}$$

Note that : If two lines have more than 1 intersection,

they have infinite intersections (i.e. they are the same line).

#11. Find the value of $a + b$ for the following systems

(1) $\begin{cases} 2x - \frac{y}{3} = \frac{2}{3} \\ (x - y):3 = -1:1 \end{cases}$ **with solution $(a, b - 1)$.**

$$\begin{cases} 6x - y = 2 \\ x - y = -3 \end{cases}$$

$$\Rightarrow \qquad 6x - y = 2$$
$$- \;) \; \underline{x - y = -3}$$
$$5x \quad = 5 \;; x = 1 \;\therefore y = x + 3 = 1 + 3 = 4$$

$\therefore (1, 4) = (a, b - 1) \;;\; a = 1, b = 5$

So, $a + b = 6$

(2) $\begin{cases} \frac{3}{x} + \frac{2}{y} = b \\ \frac{1}{x} + \frac{2}{y} = \frac{1}{2} \end{cases}$ with solution $(a, 2a)$.

Since $(a, 2a)$ is the solution, substitute it to the system. Then,

$\begin{cases} \frac{3}{a} + \frac{2}{2a} = b \\ \frac{1}{a} + \frac{2}{2a} = \frac{1}{2} \end{cases}$ $\Rightarrow$ $\begin{cases} \frac{3}{a} + \frac{1}{a} = b \\ \frac{1}{a} + \frac{1}{a} = \frac{1}{2} \end{cases}$ $\Rightarrow$ $\begin{cases} \frac{4}{a} = b \\ \frac{2}{a} = \frac{1}{2} \end{cases}$ $\therefore$ $a = 4$, $b = 1$

So, $a + b = 4 + 1 = 5$

(3) $ax + (b-1)y = 2ax - 3y + 5 = x + by - 1$ with solution $(2, 3)$.

Since $(2, 3)$ is the solution, $2a + 3(b-1) = 4a - 9 + 5 = 2 + 3b - 1$

So, $2a + 3b - 3 = 4a - 4 = 3b + 1$

$\begin{cases} 2a + 3b - 3 = 4a - 4 \\ 2a + 3b - 3 = 3b + 1 \end{cases}$ $\Rightarrow$ $\begin{cases} 2a - 3b = 1 \\ 2a = 4 \end{cases}$ $\Rightarrow$ $\begin{cases} 2a - 3b = 1 \\ a = 2 \end{cases}$

$\therefore 4 - 3b = 1$; $3b = 3$; $b = 1$

So, $a + b = 2 + 1 = 3$

#12. The system $\begin{cases} 3x - 2y = k \\ -2x + y = 3 \end{cases}$ **has the solution** (a, b) **with the condition** $a : b = 1 : 3$.

Find the constant k.

Since $a : b = 1 : 3$, $3a = b$

Since (a, b) is the solution, $\begin{cases} 3a - 2b = k \\ -2a + b = 3 \end{cases}$

$\therefore \begin{cases} 3a - 2b = k \\ -2a + b = 3 \end{cases}$ $\Rightarrow$ $\begin{cases} 3a - 6a = k \\ -2a + 3a = 3 \end{cases}$ $\Rightarrow$ $\begin{cases} -3a = k \\ a = 3 \end{cases}$

$\therefore k = -3a = -9$

#13. Find the value of $x + y$ **for variables** x **and** y **that satisfy the equations** $2^x \cdot 8^y = 32$ **and**
$3^{x+1} \cdot 9^{y-1} = 3^3$.

$2^x \cdot 8^y = 32 \Rightarrow 2^x \cdot (2^3)^y = 2^5 \Rightarrow 2^{x+3y} = 2^5$; $x + 3y = 5$

$3^{x+1} \cdot 9^{y-1} = 3^3 \Rightarrow 3^{x+1} \cdot (3^2)^{y-1} = 3^3 \Rightarrow 3^{x+1+2y-2} = 3^3$

$x + 1 + 2y - 2 = 3$; $x + 2y = 4$

$\therefore \begin{cases} x + 3y = 5 \\ x + 2y = 4 \end{cases}$

$\Rightarrow \qquad x + 3y = 5$

$\qquad - \underline{) \; x + 2y = 4}$

$\qquad\qquad y = 1$; $x = 5 - 3y = 5 - 3 = 2$

$\therefore x + y = 2 + 1 = 3$

#14. Find the value of $\frac{1}{x} - \frac{1}{y}$ for variables x and y that satisfy the system

$$\begin{cases} 2x - xy - 2y - 3 = 0 \\ 3x + 2xy - 3y + 1 = 0 \end{cases}.$$

$$\begin{cases} 2x - xy - 2y - 3 = 0 \\ 3x + 2xy - 3y + 1 = 0 \end{cases} \Rightarrow \begin{cases} 2(x - y) - xy = 3 \\ 3(x - y) + 2xy = -1 \end{cases}$$

Let $x - y = A$ and $xy = B$. Then,

$$\Rightarrow \begin{cases} 2A - B = 3 \\ 3A + 2B = -1 \end{cases} \Rightarrow \begin{cases} 4A - 2B = 6 \\ 3A + 2B = -1 \end{cases}$$

$$\Rightarrow \quad 4A - 2B = 6$$
$$+ \underline{\quad)\ 3A + 2B = -1}$$
$$7A \qquad = 5 \ ; A = \frac{5}{7}$$

$$\therefore B = 2A - 3 = \frac{10}{7} - 3 = \frac{-11}{7}$$

Since $A = \frac{5}{7}$, $x - y = \frac{5}{7}$.

Since $B = \frac{-11}{7}$, $xy = \frac{-11}{7}$.

$$\therefore \frac{1}{x} - \frac{1}{y} = \frac{y - x}{xy} = \frac{\frac{5}{7}}{\frac{-11}{7}} = \frac{5}{11}$$

OR $\begin{cases} 2(x - y) - xy = 3 \\ 3(x - y) + 2xy = -1 \end{cases} \Rightarrow \begin{cases} \frac{2(x-y)}{xy} - 1 = \frac{3}{xy} \\ \frac{3(x-y)}{xy} + 2 = \frac{-1}{xy} \end{cases} \Rightarrow \begin{cases} \frac{2(x-y)}{xy} - 1 = \frac{3}{xy} \\ \frac{9(x-y)}{xy} + 6 = \frac{-3}{xy} \end{cases}$

Adding the two equations gives

$$\frac{11(x-y)}{xy} + 5 = 0 \quad \Rightarrow \frac{(x-y)}{xy} = -\frac{5}{11} \quad \therefore \frac{1}{x} - \frac{1}{y} = \frac{y-x}{xy} = \frac{5}{11}$$

#15. The perimeter of a rectangle is 18 inches. The length of the rectangle is 3 inches shorter than twice its width. What is the area of the rectangle?

Let x = the length of a rectangle

y = the width of a rectangle. Then,

$$\begin{cases} 2(x + y) = 18 \\ x = 2y - 3 \end{cases} \Rightarrow \begin{cases} x + y = 9 \\ x = 2y - 3 \end{cases} \Rightarrow (2y - 3) + y = 9 \ ; 3y = 12 \ ; \ y = 4$$

$$\therefore x = 2y - 3 = 8 - 3 = 5$$

Therefore, the area is 20 square inches.

#16. Movie ticket prices are $6 for children and $9 for adults. Nichole pays $84 for 12 people. How many children are in her group?

Let x = the number of children

y = the number of adult

Then, $\begin{cases} x + y = 12 \\ 6x + 9y = 84 \end{cases}$ $\Rightarrow$ $\begin{cases} 6x + 6y = 72 \\ 6x + 9y = 84 \end{cases}$

$\Rightarrow \qquad 6x + 6y = 72$

$ -) \, \underline{6x + 9y = 84}$

$ 3y = 12 \quad ; y = 4 \; \therefore x = 8$

Therefore, 8 children are in her group.

#17. Apples and peaches are mixed in a box. There are 3 less apples than three times the number of peaches. Two times the total number of apples and peaches is 10. How many apples and peaches are in the box?

Let A = the number of apple

P = the number of peach

Then, $\begin{cases} A = 3P - 3 \\ 2(A + P) = 10 \end{cases}$ $\Rightarrow$ $\begin{cases} A = 3P - 3 \\ A + P = 5 \end{cases}$

$\therefore 3P - 3 + P = 5 \; ; 4P = 8 \; ; \; P = 2 \; \therefore A = 3$

Therefore, there are 3 apples and 2 peaches in the box.

#18. Richard prepares a bag of candies for kids. If each kid gets 8 candies, then 8 candies will be left. If they each get 10 candies then Richard will be short 6 candies. How many candies are in Richard's bag?

Let x = the number of candies

y = the number of kids

Then, $\begin{cases} x - 8y = 8 \\ x - 10y = -6 \end{cases}$

$\Rightarrow \qquad x - 8y = 8$

$ -) \, \underline{x - 10y = -6}$

$ 2y = 14 \quad ; y = 7 \; \therefore x = 8 + 56 = 64$

Therefore, there are 64 candies in the bag.

#19. Nichole has quarters and dimes worth \$2.55 in her purse. The number of dimes is two less than three times the number of quarters.

How many quarters and dimes are in her purse?

Let x = the number of quarters

y = the number of dimes

Then, $\begin{cases} 0.25x + 0.10y = 2.55 \\ y = 3x - 2 \end{cases}$

$\Rightarrow \begin{cases} 25x + 10y = 255 \\ y = 3x - 2 \end{cases}$ $\therefore 25x + 10(3x - 2) = 255$; $55x = 275$; $x = 5$

So, $y = 15 - 2 = 13$

Therefore, there are 5 quarters and 13 dimes.

#20. If you add the ten's digit and the one's digit of a certain two-digit integer, the sum is 12. If the digits of the number are interchanged, the new number will be 12 less than twice the original number. Find the original number.

Let x = the ten's digit of original number

y = the one's digit of original number

Then, $\begin{cases} x + y = 12 \\ 10y + x = 2(10x + y) - 12 \end{cases}$

$\Rightarrow \begin{cases} x + y = 12 \\ 19x - 8y = 12 \end{cases}$

Since $x + y = 12$, $y = 12 - x$; $19x - 8(12 - x) = 12$; $19x - 96 + 8x = 12$; $27x = 108$

$\therefore x = 4$, $y = 12 - x = 8$

So, the original number is 48.

#21. If 30 ounces of salt solution containing a x% of salt solution is added to 40 ounces of salt solution containing a y% of salt solution, it produces a salt solution that is 15% salt. If 30 ounces of a salt solution containing a y% of salt solution is added to 40 ounces of a salt solution containing a x% of salt solution, it produces a salt solution that is 18% salt. Find the values of x and y.

$$40 \cdot \frac{y}{100} + 30 \cdot \frac{x}{100} = 70 \cdot \frac{15}{100} \quad ; \quad 40y + 30x = 70 \cdot 15 \; ; \; 4y + 3x = 7 \cdot 15$$

$$40 \cdot \frac{x}{100} + 30 \cdot \frac{y}{100} = 70 \cdot \frac{18}{100} \quad ; \quad 40x + 30y = 70 \cdot 18 \; ; \; 4x + 3y = 7 \cdot 18$$

$$\Rightarrow \quad \begin{cases} 4y + 3x = 7 \cdot 15 \\ 4x + 3y = 7 \cdot 18 \end{cases} \quad \Rightarrow \quad \begin{cases} 12x + 16y = 7 \cdot 15 \cdot 4 \\ 12x + 9y = 7 \cdot 18 \cdot 3 \end{cases}$$

$$\Rightarrow \qquad 12x + 16y = 7 \cdot 15 \cdot 4$$
$$-) \; \underline{12x + 9y = 7 \cdot 18 \cdot 3}$$
$$7y = 7(60 - 54) \; ; \; y = 6$$
$$\therefore \; 3x = 7 \cdot 15 - 4y = 105 - 24 = 81 \; ; \; x = 27$$

Therefore, $x = 27, \; y = 6$

#22. Richard wants to produce 70 ounces of salt solution that is 8% salt by adding water after mixing salt solution that is 5% salt with salt solution that is 10% salt. The amount of a 10% of salt solution is three times as much as the amount of a 5% of salt solution. How much water should he add?

Note that water is a 0% of salt solution.

Let $x =$ amount of salt solution containing a 5% of salt solution

$y =$ amount of water to be added.

Then, $x + 3x + y = 70 \; ; \; 4x + y = 70$

$x \cdot \frac{5}{100} + 3x \cdot \frac{10}{100} + y \cdot \frac{0}{100} = 70 \cdot \frac{8}{100}$ ($\because$ same amount of salt)

$5x + 30x = 70 \cdot 8 \; ; \quad 35x = 70 \cdot 8 \; ; \; x = 16$

Since $4x + y = 70, \; y = 70 - 4x = 70 - 4 \cdot 16 = 6$

Therefore, Richard should add 6 ounces of water.

#23. Nichole wants to produce 29 liters of 20% alcohol solution by adding alcohol after mixing two alcohol solutions that are 4% alcohol and 3% alcohol separately. The amount of alcohol solution that is 3% alcohol is twice the amount of alcohol solution that is 4% alcohol. How many liters of alcohol must be added?

Let x = amount of alcohol solution containing 4% alcohol

y = amount of alcohol to be added.

Then, $2x + x + y = 29$; $3x + y = 29$

$2x \cdot \frac{3}{100} + x \cdot \frac{4}{100} + y \cdot \frac{100}{100} = 29 \cdot \frac{20}{100}$

$\therefore$ $6x + 4x + 100y = 20 \cdot 29$; $10x + 100y = 580$; $x + 10y = 58 = 2 \cdot 29$

$\begin{cases} 3x + 30y = 6 \cdot 29 \\ 3x + y = 29 \end{cases}$

$\Rightarrow$ $\quad 3x + 30y = 6 \cdot 29$

$-)\ 3x + y = 29$

$\quad\quad\quad 29y = 29(6 - 1)$; $y = 5$

Therefore, she needs to add 5 liters of alcohol.

#24. Nichole started to run at 9:50AM with a speed of 8 miles per hour and then walked the rest of the way at 3 miles per hour. She arrived at 10:40AM. If Nichole went to the park which was 4 miles away, then how many miles did she run?

Let x be the distance she ran and y be the distance she walked. Then,

$\begin{cases} x + y = 4 \\ \frac{x}{8} + \frac{y}{3} = \frac{50}{60} \end{cases}$ $\Rightarrow$ $\begin{cases} 3x + 3y = 12 \\ 3x + 8y = 20 \end{cases}$

$\Rightarrow$ $\quad 3x + 3y = 12$

$-)\ 3x + 8y = 20$

$\quad\quad\quad 5y = 8$; $y = \frac{8}{5} = 1.6$ $\quad \therefore x = 4 - y = 4 - 1.6 = 2.4$

Therefore, she ran for 2.4 miles.

#25. Richard starts a trail ride in a parking lot. He rides up a long hill on *A* trail at 4 miles per hour and comes down the hill on *B* trail at 12 miles per hour. His ride takes 1 hour 20 minutes total. The total distance of *A* trail and *B* trail is 10 miles. How many miles long is *B* trail?

Let $x =$ the distance for the *A* trail

$y =$ the distance for the *B* trail

Then, $\begin{cases} x + y = 10 \\ \frac{x}{4} + \frac{y}{12} = 1\frac{20}{60} \end{cases} \Rightarrow \begin{cases} x + y = 10 \\ 3x + y = 16 \end{cases}$

$\Rightarrow \quad 3x + y = 16$

$\underline{-)\ x + y = 10}$

$\quad 2x \qquad = 6 \ ; \ x = 3 \ \therefore y = 7$

Therefore, *B* trail is 7 miles long.

#26.   Richard and Nichole want to finish a job. Richard works alone for 3 hours and leaves. Nichole comes and works alone the rest of the job for another 3 hours, thereby finishing the job. OR if Richard works alone for 6 hours and then Nichole works alone for 2 hours for the rest of the job after Richard is done, the job is also completed. How long will it take Richard to finish the job by himself the entire time?

Note that $\dfrac{\text{Richard's time worked}}{\text{Richard's time to complete}} + \dfrac{\text{Nichole's time worked}}{\text{Nichole's time to complete}} = 1$

, where 1 represents complete the job.

Let x be the Richard's time to complete the job alone

y be the Nicholes's time to complete the job alone

$\Rightarrow \begin{cases} \frac{3}{x} + \frac{3}{y} = 1 \\ \frac{6}{x} + \frac{2}{y} = 1 \end{cases} \Rightarrow \begin{cases} 3y + 3x = xy \\ 6y + 2x = xy \end{cases}$

$\therefore \ 3y + 3x = 6y + 2x \ ; \ x = 3y$

Since $\frac{3}{x} + \frac{3}{y} = 1, \ \frac{3}{3y} + \frac{3}{y} = 1 \ \therefore \frac{1}{y} + \frac{3}{y} = 1 \ ; \ \frac{4}{y} = 1 \ ; \ y = 4 \ \therefore x = 12$

Therefore, Richard will take 12 hours to finish the job alone.

#27. 5 years ago, Nichole was 5 years less than one-third her mom's age. In 6 years, her mom will be 10 years more than twice Nichole's age at that time. How old was Nichole's mom when Nichole was 15?

Let x = Nichole's current age

y = mom's current age.

Then, $\begin{cases} x - 5 = \frac{1}{3}(y - 5) - 5 \\ y + 6 = 2(x + 6) + 10 \end{cases}$ $\Rightarrow$ $\begin{cases} 3x - 15 = y - 5 - 15 \\ y - 2x = 12 + 10 - 6 \end{cases}$ $\Rightarrow$ $\begin{cases} 3x - y = -5 \\ y - 2x = 16 \end{cases}$

$\Rightarrow \quad 3x - y = -5$

$+)\underline{-2x + y = 16}$

$\quad\quad x \quad\quad = 11 \; ; \; y = 3x + 5 = 38$

So, Nichole is 11 years old now .   Nichole's mom will be 42 years old when Nichole is15.

#28. Richard walks from home to a library at 3 miles per hour. 20 minutes after Richard leaves, Nichole rides a bike at 8 miles per hour from home to the library. They arrive at the same time. How long does it take Richard to meet Nichole at the library?

The distance that both Richard and Nichole traveled to meet each other is the same.

Let t = the time for Nichole to meet Richard after she starts biking.

Then, the distance Nichole rides her bike is $8 \cdot t$ and the distance Richard walks is $3 \cdot \left(t + \frac{1}{3}\right)$.

So, $8t = 3\left(t + \frac{1}{3}\right) = 3t + 1 \; ; \; 5t = 1 \; ; \; t = \frac{1}{5}$

Thus, Nichole takes $\frac{1}{5} \cdot 60 = 12$ minutes.

Since $t + \frac{1}{3} = \frac{1}{5} + \frac{1}{3} = \frac{8}{15}$, Richard takes $\frac{8}{15} \cdot 60 = 32$ minutes to meet Nichole.

#29. If Richard drives a car from home to the doctor's office at 50 miles per hour, he will arrive at the office 5 minutes earlier than his appointment time. If he drives a car at 40 miles per hour on the same route, he will arrive 10 minutes late to his appointment. How far is the office from Richard's home?

Let t be the time for Richard to arrive at the office on time.

Then, $50\left(t - \frac{5}{60}\right) = 40\left(t + \frac{10}{60}\right)$ ($\because$ the same distance)

So, $5\left(t - \frac{1}{12}\right) = 4\left(t + \frac{1}{6}\right) \; ; \; t = \frac{4}{6} + \frac{5}{12} = \frac{8+5}{12} = \frac{13}{12} = 1\frac{1}{12}$

So, Richard needs1 hour 5 minutes to arrive at the office on time.

Since $40\left(t + \frac{1}{6}\right) = 40\left(\frac{13}{12} + \frac{1}{6}\right) = 40 \cdot \frac{15}{12} = 50$,

the distance from home to the doctor's office is 50 miles.

#30. Richard and Nichole jog towards each other from two opposite starting points 1 mile apart. Nichole jogs 1.5 times faster than Richard. How fast does Nichole have to jog if they meet each other in 30 minutes?

Let $v_1 = $ Richard's speed

$v_2 = $ Nichole's speed.

Then, $v_2 = 1.5 \times v_1$

Since $v_1 \cdot \frac{1}{2} + v_2 \cdot \frac{1}{2} = 1$, $v_1 \cdot \frac{1}{2} + (1.5\, v_1) \cdot \frac{1}{2} = 1$

$v_1 + 1.5\, v_1 = 2$; $2.5\, v_1 = 2$; $v_1 = \frac{2}{2.5} = 0.8$ mph.

$\therefore v_2 = 1.5 \times 0.8 = 1.2$ mph

Therefore, Nichole must jog at 1.2 miles per hour.

#31. There were 44 boys and girls in a math club last summer. This year, 25% of the boys quit and 15% of the girls joined the club again. Now the club has 41 members. How many boys and girls are in the club now?

Let $x = $ the number of boys, last summer

$y = $ the number of girls, last summer

Then $\begin{cases} x + y = 44 \\ -\frac{25}{100}x + \frac{15}{100}y = -3 \end{cases}$

$\Rightarrow \begin{cases} x + y = 44 \\ -25x + 15y = -300 \end{cases} \Rightarrow \begin{cases} 15x + 15y = 660 \\ -25x + 15y = -300 \end{cases}$

$\Rightarrow \quad 15x + 15y = 660$

$-)\,-25x + 15y = -300$

$\overline{}$

$40x \quad\quad = 960$; $4x = 96$; $x = 24 \quad \therefore y = 20$

Therefore, the current number of boys in the club is $24 - \frac{25}{100} \cdot 24 = 18$ and

the current number of girls in the club is $20 + \frac{15}{100} \cdot 20 = 23$.

#32. Six years ago, Nichole was three times as old as Richard. Four years from now, Nichole will be twice as old as Richard. How old is Richard now?

Let N be the current age of Nichole and R be the current age of Richard. Then,

$$\begin{cases} N - 6 = 3(R - 6) \\ N + 4 = 2(R + 4) \end{cases} \Rightarrow \begin{cases} N - 3R = -12 \\ N - 2R = 4 \end{cases}$$

$$\Rightarrow \quad N - 2R = 4$$
$$-)\ \underline{N - 3R = -12}$$
$$R = 16$$

Therefore, Richard is 16 years old now.

#33. Solve each system by graphing.

(1) $\begin{cases} x + y = 4 \\ 2x - y = 5 \end{cases}$

$$\Rightarrow \begin{cases} y = -x + 4 \\ y = 2x - 5 \end{cases}$$

$y = -x + 4 \Rightarrow$ x-intercept is $(4, 0)$ and y-intercept is $(0, 4)$.

$y = 2x - 5 \Rightarrow$ x-intercept is $\left(\frac{5}{2}, 0\right)$ and y-intercept is $(0, -5)$.

∴ Solution is $(x, y) = (3, 1)$

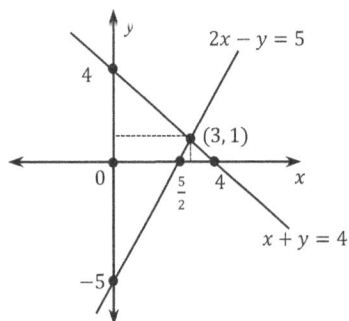

(2) $\begin{cases} 2x - y = 5 \\ 4x - 2y = 6 \end{cases}$

$\Rightarrow \begin{cases} y = 2x - 5 \\ y = 2x - 3 \end{cases}$

$y = 2x - 5 \Rightarrow$ x-intercept is $\left(\frac{5}{2}, 0\right)$ and y-intercept is $(0, -5)$.

$y = 2x - 3 \Rightarrow$ x-intercept is $\left(\frac{3}{2}, 0\right)$ and y-intercept is $(0, -3)$.

By graphing, the two equations are parallel. So, there is no intersection point.

Therefore, the system has no solution.

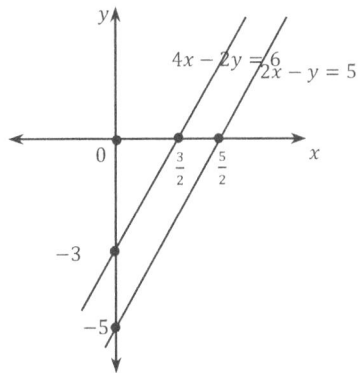

(3) $\begin{cases} 3x + 4y = -5 \\ 6x + 8y = -10 \end{cases}$

$\Rightarrow \begin{cases} y = -\frac{3}{4}x - \frac{5}{4} \\ y = -\frac{6}{8}x - \frac{10}{8} \end{cases}$

$y = -\frac{3}{4}x - \frac{5}{4} \Rightarrow$ x-intercept is $\left(-\frac{5}{3}, 0\right)$ and y-intercept is $\left(0, -\frac{5}{4}\right)$

$y = -\frac{6}{8}x - \frac{10}{8} \Rightarrow$ x-intercept is $\left(-\frac{5}{3}, 0\right)$ and y-intercept is $\left(0, -\frac{5}{4}\right)$

By graphing, the two equations are coinciding.

So, the system has unlimited number of solutions.

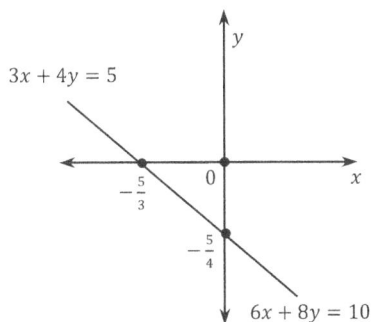

Solutions for Chapter 5.

#1. Find the range of x for the following systems

(1) $\begin{cases} x \geq -2 \\ x < 4 \end{cases}$ $-2 \leq x < 4$

(2) $\begin{cases} x > -3 \\ x > 0 \end{cases}$ $x > 0$

(3) $\begin{cases} x < -1 \\ x \leq 2 \end{cases}$ $x < -1$

(4) $\begin{cases} x \leq 1 \\ x \geq 3 \end{cases}$ no solution

(5) $\begin{cases} x \geq 2 \\ x < 2 \end{cases}$ no solution

(6) $\begin{cases} x \leq -3 \\ x \geq -3 \end{cases}$ $x = -3$

#2. Solve the following systems

(1) $\begin{cases} 3x - 1 < 4x + 2 \\ x + 3 \geq -2x \end{cases}$ $\Rightarrow$ $\begin{cases} x > -3 \\ 3x \geq -3 \; ; x \geq -1 \end{cases}$

$\therefore \; x \geq -1$

(2) $\begin{cases} -2x + 4 > 3x - 6 \\ 3x - 5 \geq x + 3 \end{cases}$ $\Rightarrow$ $\begin{cases} 5x < 10 \; ; x < 2 \\ 2x \geq 8 \; ; x \geq 4 \end{cases}$

$\therefore$ no solution

(3) $\begin{cases} x - 2 > -5 \\ 5x - 3 < 2x + 3 \end{cases}$ $\Rightarrow$ $\begin{cases} x > -3 \\ 3x < 6 \; ; x < 2 \end{cases}$

$\therefore -3 < x < 2$

(4) $\begin{cases} 4x - 3 > 5x - 2 \\ 2x - 2 > 3x - 5 \end{cases}$ $\Rightarrow \begin{cases} x < -1 \\ x < 3 \end{cases}$

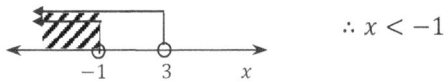

$\therefore x < -1$

(5) $\begin{cases} \frac{x+2}{2} \leq \frac{x-2}{3} + x \\ 2(x-2) - \frac{x}{2} > 2x - 6 \end{cases}$ $\Rightarrow \begin{cases} 3(x+2) \leq 2(x-2) + 6x \\ 2x - 4 - \frac{x}{2} > 2x - 6 \end{cases}$ $\Rightarrow \begin{cases} 5x \geq 10 \\ \frac{x}{2} < 2 \end{cases}$ $\Rightarrow \begin{cases} x \geq 2 \\ x < 4 \end{cases}$

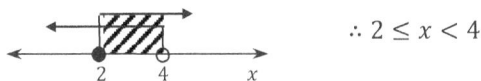

$\therefore 2 \leq x < 4$

(6) $x - 3 < \frac{x}{2} + 1 \leq 3x - 1$ $\Rightarrow \begin{cases} x - 3 < \frac{x}{2} + 1 \\ \frac{x}{2} + 1 \leq 3x - 1 \end{cases}$ $\Rightarrow \begin{cases} \frac{x}{2} < 4 \\ \frac{5x}{2} \geq 2 \end{cases}$ $\Rightarrow \begin{cases} x < 8 \\ x \geq \frac{4}{5} \end{cases}$

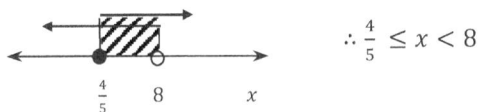

$\therefore \frac{4}{5} \leq x < 8$

(7) $0.5x + \frac{x}{2} + 1 > 0.7x - 0.2 > 0.3x - 0.6$

$\Rightarrow \begin{cases} 0.5x + \frac{x}{2} + 1 > 0.7x - 0.2 \\ 0.7x - 0.2 > 0.3x - 0.6 \end{cases}$ $\Rightarrow \begin{cases} 5x + 5x + 10 > 7x - 2 \\ 7x - 2 > 3x - 6 \end{cases}$ $\Rightarrow \begin{cases} 3x > -12 \\ 4x > -4 \end{cases}$ $\Rightarrow \begin{cases} x > -4 \\ x > -1 \end{cases}$

$\therefore x > -1$

(8) $\begin{cases} 2.1x > 3.6 - 0.9x \\ 3(x-1) - 2 < \frac{1}{2}(3x+1) \end{cases}$ $\Rightarrow \begin{cases} 21x > 36 - 9x \\ 3x - 5 < \frac{3}{2}x + \frac{1}{2} \end{cases}$ $\Rightarrow \begin{cases} 30x > 36 \\ \frac{3}{2}x < \frac{11}{2} \end{cases}$ $\Rightarrow \begin{cases} x > \frac{6}{5} \\ x < \frac{11}{3} \end{cases}$

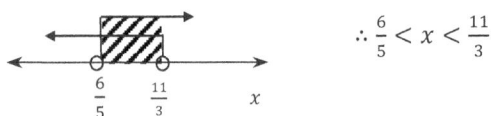

$\therefore \frac{6}{5} < x < \frac{11}{3}$

#3. Find the range of $y = -2x + 3$ when x satisfies the following systems

(1) $\begin{cases} x - 3 \leq 2 \\ -3x + 2 < 4 \end{cases} \Rightarrow \begin{cases} x \leq 5 \\ 3x > -2 \end{cases}$

$\therefore -\frac{2}{3} < x \leq 5 \; ; \; \frac{4}{3} > -2x \geq -10 \; ; \; 4\frac{1}{3} > -2x + 3 \geq -7$

$\therefore -7 \leq y < 4\frac{1}{3}$

(2) $\begin{cases} 2x - 3 < 4x - 1 \\ 5x - 2 \geq 3x + 2 \end{cases} \Rightarrow \begin{cases} 2x > -2 \\ 2x \geq 4 \end{cases} \Rightarrow \begin{cases} x > -1 \\ x \geq 2 \end{cases}$

$\therefore x \geq 2 \; ; \; -2x \leq -4 \; ; \; -2x + 3 \leq -4 + 3$

$\therefore y \leq -1$

(3) $\frac{x-1}{3} \leq \frac{1}{4}(x - 3) < \frac{5-3x}{4} \Rightarrow \begin{cases} \frac{x-1}{3} \leq \frac{1}{4}(x - 3) \\ \frac{1}{4}(x - 3) < \frac{5-3x}{4} \end{cases} \Rightarrow \begin{cases} 4(x - 1) \leq 3(x - 3) \\ (x - 3) < 5 - 3x \end{cases} \Rightarrow \begin{cases} x \leq -5 \\ x < 2 \end{cases}$

$\therefore x \leq -5 \; ; \; -2x \geq 10 \; ; \; -2x + 3 \geq 10 + 3$

$\therefore y \geq 13$

#4. Find the value of k for the following conditions

(1) The system $\begin{cases} x + 5 < 2k \\ 3x - 2 \geq 1 \end{cases}$ has the solution $2 \leq x < 5$.

$\Rightarrow \begin{cases} x < 2k - 5 \\ 3x \geq 6 \end{cases} \Rightarrow \begin{cases} x < 2k - 5 \\ x \geq 2 \end{cases}$

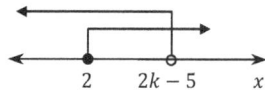

$\therefore 2 \leq x < 2k - 5$

$2k - 5 = 5 \; ; \; 2k = 10 \quad \therefore k = 5$

(2) The system $\begin{cases} \frac{2x+1}{3} > \frac{x-3}{5} \\ 0.6x - 2.4 < kx - 0.8 \end{cases}$ has the solution $-2 < x < 2$.

$\Rightarrow \begin{cases} 5(2x + 1) > 3(x - 3) \\ 6x - 24 < 10kx - 8 \end{cases} \Rightarrow \begin{cases} 7x > -14 \\ (6 - 10k)x < 16 \end{cases} \Rightarrow \begin{cases} x > -2 \\ x < \frac{16}{6-10k} \end{cases}$

$\therefore -2 < x < \frac{16}{6-10k}$

$\frac{16}{6-10k} = 2 \; ; \; 12 - 20k = 16 \; ; \; 20k = -4$

$\therefore k = -\frac{1}{5}$

(3) The system $\begin{cases} -x + 2 \leq 0 \\ \frac{x}{2} + 3 \leq -k + 5 \end{cases}$ has the solution $x = 2$.

$\Rightarrow \begin{cases} x \geq 2 \\ \frac{x}{2} \leq -k + 2 \end{cases} \Rightarrow \begin{cases} x \geq 2 \\ x \leq 2(-k + 2) \end{cases}$

$\therefore 2(-k + 2) = 2 \ ; \ -k + 2 = 1 \quad \therefore \ k = 1$

(4) The system $\begin{cases} \frac{k-x}{2} \leq x + 5 \\ 3 - 2x < 3x - 2 \end{cases}$ has the solution $x \geq 3$.

$\Rightarrow \begin{cases} k - x \leq 2x + 10 \\ 3 - 2x < 3x - 2 \end{cases} \Rightarrow \begin{cases} 3x \geq k - 10 \\ 5x > 5 \end{cases} \Rightarrow \begin{cases} x \geq \frac{k-10}{3} \\ x > 1 \end{cases}$

$\therefore \dfrac{k - 10}{3} = 3$

$\therefore k = 19$

(5) The system $\begin{cases} 2x + 3 \leq 4x - 5 \\ 3(x - k) \leq x + 3 \end{cases}$ has only one solution.

$\Rightarrow \begin{cases} 2x \geq 8 \\ 2x \leq 3 + 3k \end{cases} \Rightarrow \begin{cases} x \geq 4 \\ x \leq \frac{3+3k}{2} \end{cases}$

$\therefore \dfrac{3 + 3k}{2} = 4 \ ; \ 3 + 3k = 8 \ ; \ 3k = 5$

$\therefore k = \dfrac{5}{3}$

(6) The system $3 < \dfrac{k-4x}{-2} < 5$ has the solution $1 < x < 2$.

$3 < \frac{k-4x}{-2} < 5 \Rightarrow -6 > k - 4x > -10 \Rightarrow -6 - k > -4x > -10 - k \ ; \ \frac{-6-k}{-4} < x < \frac{-10-k}{-4}$

$\therefore \frac{-6-k}{-4} = 1$ and $\frac{-10-k}{-4} = 2$

$\therefore$ if $\frac{-6-k}{-4} = 1 \ \Rightarrow -6 - k = -4 \ ; \ k = -2$ and if $\frac{-10-k}{-4} = 2 \Rightarrow -10 - k = -8 \ ; \ k = -2$

Therefore, $k = -2$

#5. Find the range of k for the following conditions

(1) The system $\begin{cases} 2x \le 5 - k \\ 3x - 3 \ge 2x - 1 \end{cases}$ **has no solution.**

$\Rightarrow \begin{cases} x \le \dfrac{5-k}{2} \\ x \ge 2 \end{cases}$

$\therefore \dfrac{5-k}{2} < 2 \ ; \ 5 - k < 4 \ ; \ k > 1$

(2) The system $\begin{cases} x - 3 \le 2x - 6 \\ 5x + k < 3x + 1 \end{cases}$ **has no solution.**

$\Rightarrow \begin{cases} x \ge 3 \\ 2x < 1 - k \end{cases} \Rightarrow \begin{cases} x \ge 3 \\ x < \dfrac{1-k}{2} \end{cases}$

If $\begin{cases} x \ge 3 \\ x < 3 \end{cases}$, then there is no solution. Therefore, $\dfrac{1-k}{2} \le 3 \ ; \ 1 - k \le 6$

$\therefore k \ge -5$

(3) The system $\begin{cases} 2r + 3 \le -5 \\ x + k > 1 \end{cases}$ **has only one integer in the solution.**

$\Rightarrow \begin{cases} 2x \le -8 \\ x > 1 - k \end{cases} \Rightarrow \begin{cases} x \le -4 \\ x > 1 - k \end{cases}$

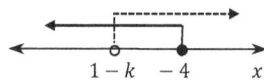

If $1 - k = -4$, then $\begin{cases} x \le -4 \\ x > 4 \end{cases}$; no solution.

If $1 - k > -4$, then there is no solution.

If $1 - k = -5$, then $\begin{cases} x \le -4 \\ x > -5 \end{cases}$; $x = -4$ is the only one integer.

$\therefore -5 \le 1 - k < -4 \ ; \ -6 \le -k < -5 \ ; \ 6 \ge k > 5$

$\therefore 5 < k \le 6$

(4) The system $\begin{cases} x - 3 \geq 0 \\ 3x + k \leq 2x + 3 \end{cases}$ has solutions.

$\Rightarrow \begin{cases} x \geq 3 \\ x \leq 3 - k \end{cases}$

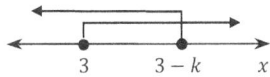

$\therefore\ 3 \leq 3 - k\ ;\ k \leq 0$

#6. Find the sum of all integers that satisfy the following systems

(1) $\begin{cases} 3x - 5 \leq 7 \\ \frac{x-1}{2} < x + 3 \\ 2x - 5 < 5x + 4 \end{cases}$

$\Rightarrow \begin{cases} 3x \leq 12 \\ x - 1 < 2x + 6 \\ 3x > -9 \end{cases} \Rightarrow \begin{cases} x \leq 4 \\ x > -7 \\ x > -3 \end{cases}$

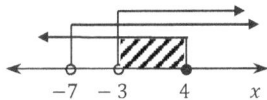

$\therefore$ The integers are $-2, -1, 0, 1, 2, 3,$ and 4.

$\therefore$ The sum is $-2 + (-1) + 0 + 1 + 2 + 3 + 4 = 7$.

(2) $\begin{cases} |x| \leq 5 \\ |x| > 2 \end{cases}$

$\Rightarrow \begin{cases} -5 \leq x \leq 5 \\ x > 2 \text{ or } x < -2 \end{cases}$

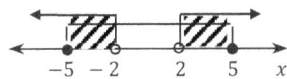

$\therefore\ -5 \leq x < -2\ $ or $\ 2 < x \leq 5$

$\therefore$ The integers are $-5, -4, -3,\ 3,\ 4,\ 5$

$\therefore$ The sum is $-5 + (-4) + (-3) + 3 + 4 + 5 = 0$.

#7. The system $\begin{cases} x - 1 \geq 2x - 4 \\ \frac{x+k}{2} < 3x - 2 \end{cases}$ **has 5 integers in the solution. What is the minimum value for k?**

$\Rightarrow \begin{cases} x \leq 3 \\ x + k < 6x - 4 \end{cases} \Rightarrow \begin{cases} x \leq 3 \\ 5x > k + 4 \end{cases} \Rightarrow \begin{cases} x \leq 3 \\ x > \frac{k+4}{5} \end{cases}$

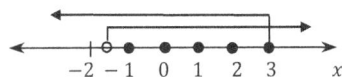

$\therefore\ -2 \leq \dfrac{k+4}{5} < -1\ ;\ -10 \leq k + 4 < -5\ ;\ -14 \leq k < -9$

$\therefore$ The minimum value for k is -14.

#8. **3 more than twice a number is less than or equal to 8, and −1 is less than one fourth of the number.**

(1) **Find the range of the number.**

$$\Rightarrow \begin{cases} 2x + 3 \le 8 \\ -1 < \frac{1}{4}x \end{cases} \Rightarrow \begin{cases} 2x \le 5 \\ x > -4 \end{cases} \Rightarrow \begin{cases} x \le \frac{5}{2} \\ x > -4 \end{cases}$$

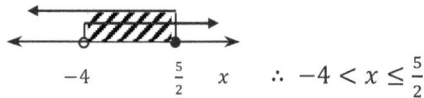

$$\therefore -4 < x \le \frac{5}{2}$$

(2) **Find the sum of the maximum integer and minimum integer that satisfies the system.**

Since $-4 < x \le \frac{5}{2}$, the maximum integer is 2 and the minimum integer is -3.

∴ The sum is $2 + (-3) = -1$.

(3) **Solve the system with the condition, −1 is greater than one fourth of the number, instead of the second inequality in the system shown in (1). Find the maximum integer that satisfies the new system.**

$$\Rightarrow \begin{cases} 2x + 3 \le 8 \\ -1 > \frac{1}{4}x \end{cases}$$

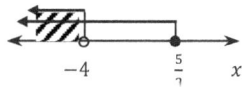

∴ $x < -4$

∴ The maximum integer is -5.

#9. **The sum of three consecutive positive integers is greater than 60 and less than 65.**

Find the largest number.

$$60 < x + (x + 1) + (x + 2) < 65$$

$$\Rightarrow 60 < 3x + 3 < 65 \quad \Rightarrow 57 < 3x < 62 \; ; \; \frac{57}{3} < x < \frac{62}{3}$$

∴ $x = 20$ is the largest integer that satisfies the system.

Therefore, the largest number among the three consecutive integers is $x + 2 = 22$.

Or, $60 < (x - 1) + x + (x + 1) < 65$

$$\Rightarrow 60 < 3x < 65 \quad \Rightarrow 20 < x < \frac{65}{3} \quad \therefore x = 21$$

Therefore, the largest number is $x + 1 = 22$.

#10. The lengths of three sides of a triangle are $x - 4$, $x + 1$, and $x + 3$. Find the range of x.

$$\Rightarrow \begin{cases} x - 4 > 0 \\ x + 3 < (x + 1) + (x - 4) \end{cases}$$

∴ The shortest length must be positive.

∴ $x + 3$ is the longest length.

by the "triangle inequality"

$$\Rightarrow \begin{cases} x > 4 \\ x > 6 \end{cases}$$

∴ $x > 6$

#11. Nichole wants to produce new salt solution by adding salt into 20 ounces of a 15% salt solution. It will be a salt concentration greater than that of a 20% salt solution and less than that of a 25% salt solution. How much salt does she need to add?

Let x be the ounces of salt Nichole will add.

$$(20 + x) \cdot \frac{20}{100} < 20 \cdot \frac{15}{100} + x \cdot \frac{100}{100} < (20 + x) \cdot \frac{25}{100}$$

$$\Rightarrow (20 + x) \cdot 20 < 20 \cdot 15 + x \cdot 100 < (20 + x) \cdot 25$$

$$\Rightarrow 400 + 20x < 300 + 100x < 500 + 25x$$

$$\Rightarrow \begin{cases} 400 + 20x < 300 + 100x \\ 300 + 100x < 500 + 25x \end{cases}$$

$$\Rightarrow \begin{cases} 100 < 80x \\ 75x < 200 \end{cases}$$

$$\Rightarrow \begin{cases} x > \frac{5}{4} \\ x < \frac{8}{3} \end{cases}$$

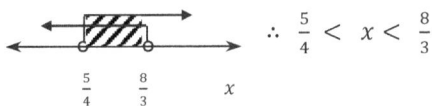

∴ $\frac{5}{4} < x < \frac{8}{3}$

#12. Richard jogs 12 miles. He begins by walking at a speed of 5 miles per hour. He then runs at a speed of 10 miles per hour for the remaining distance. If he wants to take at least 1 hour 45 minutes and at most 2 hours to complete his route, what is the longest distance he should walk?

Let x be the distance he walks. Then,

$$1\frac{3}{4} \leq \frac{x}{5} + \frac{12 - x}{10} \leq 2 \quad ; \quad \frac{7}{4} \leq \frac{2x + 12 - x}{10} \leq 2 \quad ; \quad 35 \leq 2(x + 12) \leq 40 \quad ; \quad 11 \leq 2x \leq 16$$

∴ $5\frac{1}{2} \leq x \leq 8$

Therefore, he can walk 8 miles at most.

#13. Nichole wants to reorganize all the books in her bookshelf. If she puts 30 books in each shelf, then 5 books will be left over. If she puts 35 books in each shelf, then there will be at least 20 books, but less than 25 books on the last shelf.   How many shelves are in the bookshelf?

Let x be the total number of shelves in the bookshelf. Then,

$$35(x - 1) + 20 \le 30x + 5 < 35(x - 1) + 25$$

$$\Rightarrow \begin{cases} 35(x - 1) + 20 \le 30x + 5 \\ 30x + 5 < 35(x - 1) + 25 \end{cases} \Rightarrow \begin{cases} 35x - 15 \le 30x + 5 \\ 30x + 5 < 35x - 10 \end{cases} \Rightarrow \begin{cases} 5x \le 20 \\ 5x > 15 \end{cases} \Rightarrow \begin{cases} x \le 4 \\ x > 3 \end{cases}$$

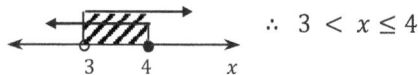 $\therefore \ 3 < x \le 4$

Since x is a positive integer, $x = 4$.

#14. Solve the following inequalities

(1) $|x - 3| < 4$

Case 1 $x - 3 \ge 0 \ (x \ge 3)$

$\Rightarrow |x - 3| = x - 3 < 4 \ ; x < 7$

Since $x \ge 3$, $3 \le x < 7$

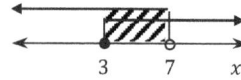

Case 2 $x - 3 < 0 \ (x < 3)$

$\Rightarrow |x - 3| = -(x - 3) < 4 \ ; x > -1$

Since $x < 3$, $-1 < x < 3$

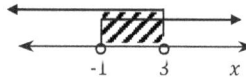

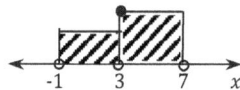

Therefore, the sum of all intervals is $-1 < x < 7$.

(2) $|x + 2| < 3x - 4$

Case 1 $x + 2 \ge 0 \ (x \ge -2)$

$\Rightarrow |x + 2| = x + 2 < 3x - 4 \ ; 2x > 6 \ ; \ x > 3$

Since $x \ge -2$, $x > 3$

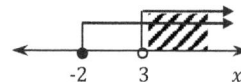

Case 2 $x + 2 < 0 \ (x < -2)$

$\Rightarrow |x + 2| = -(x + 2) < 3x - 4 \ ; 4x > 2 \ ; x > \frac{1}{2}$

Since $x < -2$, $x > \frac{1}{2}$ is not a solution.

Therefore, $x > 3$

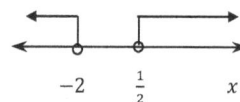

(3) $|x + 2| + |3 - x| > 10$

Since $(x + 2 = 0 \Rightarrow x = -2)$ and $(3 - x = 0 \Rightarrow x = 3)$,

consider $x < -2, -2 \leq x < 3$ and $x \geq 3$

Case 1 $x < -2$

$\Rightarrow -(x + 2) + (3 - x) > 10$; $-2x > 9$; $x < -\dfrac{9}{2}$

Since $x < -2$, $x < -\dfrac{9}{2}$

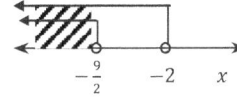

Case 2 $-2 \leq x < 3$

$\Rightarrow (x + 2) + (3 - x) > 10$; $0 \cdot x > 5$; false.

Case 3 $x \geq 3$

$\Rightarrow (x + 2) - (3 - x) > 10$; $2x > 11$; $x > \dfrac{11}{2}$

Since $x \geq 3$, $x > \dfrac{11}{2}$

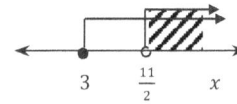

Therefore, the sum of all intervals is $x < -\dfrac{9}{2}$ or $x > \dfrac{11}{2}$.

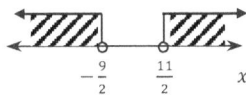

#15 Graph the following systems of linear inequalities :

(1) $\begin{cases} y < x \\ y \geq -x \end{cases}$

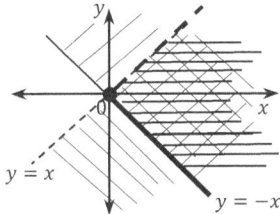

(2) $\begin{cases} 2x + y \leq 4 \\ x - y \geq 2 \end{cases} \quad \Rightarrow \quad \begin{cases} y \leq -2x + 4 \\ y \leq x - 2 \end{cases}$

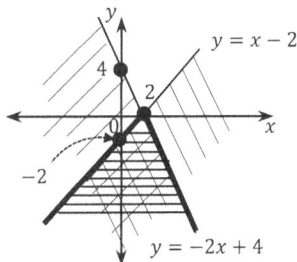

(3) $\begin{cases} 3x + y \le 6 \\ 2x + 3y > 4 \end{cases}$ $\Rightarrow$ $\begin{cases} y \le -3x + 6 \\ y > -\frac{2}{3}x + \frac{4}{3} \end{cases}$

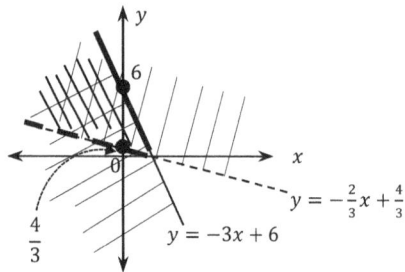

$y = -\frac{2}{3}x + \frac{4}{3}$

$\frac{4}{3}$

$y = -3x + 6$

(4) $\begin{cases} y \le 2 \\ x + y > 3 \\ x > -4 \end{cases}$ $\Rightarrow$ $\begin{cases} y \le 2 \\ y > -x + 3 \\ x > -4 \end{cases}$

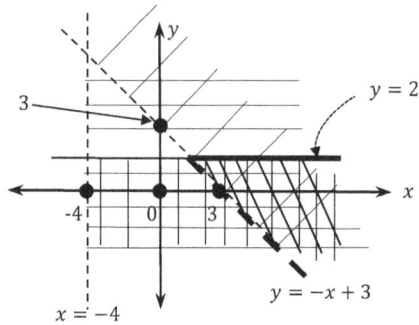

$y = 2$

$y = -x + 3$

$x = -4$

Solutions for Chapter 6

#1. Find all factors for the following expressions

(1) $ab + a + b + 1$

Since $ab + a + b + 1 = (a+1)(b+1)$, factors are $1,\ a+1,\ b+1,\ (a+1)(b+1)$.

(2) $abx + aby$

Since $abx + aby = ab(x+y)$, factors are $1,\ a,\ b,\ (x+y),\ ab,\ a(x+y),\ b(x+y),\ ab(x+y)$.

(3) $2x^2 + xy$

Since $2x^2 + xy = x(2x+y)$, factors are $1,\ x,\ 2x+y,\ x(2x+y)$.

#2. Factor the following polynomials

(1) $a^2 - ab + a = a(a - b + 1)$

(2) $a^2 b - ab^2 = ab(a - b)$

(3) $4a - 10 = 2(2a - 5)$

(4) $2a^3 + 3a^2 - 5a = a(2a^2 + 3a - 5) = a(2a + 5)(a - 1)$

(5) $9a^4 b^2 - 3a^3 b^2 + 12a^2 b^2 = 3a^2 b^2 (3a^2 - a + 4)$

(6) $2a(x + y) - 4b(x + y) = 2(x + y)(a - 2b)$

(7) $-12(x - 2y) - 18a(x - 2y) = -6(x - 2y)(2 + 3a)$

(8) $4a(a - b) + 4b(b - a) = 4(a - b)(a - b) \quad \because b - a = -(a - b)$

(9) $a^n + a^{n+2} = a^n(1 + a^2) \quad \because a^{n+2} = a^n \cdot a^2$

(10) $3x^{2n} + 12x^{3n} + 9x^n = 3x^n(x^n + 4x^{2n} + 3) \quad \because x^{2n} = x^n \cdot x^n$ and $x^{3n} = x^{2n} \cdot x^n$

#3. Factor each polynomial using factorization formulas.

(1) $x^2 - 2x + 1 = x^2 - 2 \cdot x \cdot 1 + 1^2 = (x - 1)^2$

(2) $9x^2 + 6x + 1 = (3x)^2 + 2 \cdot 3x \cdot 1 + 1^2 = (3x + 1)^2$

(3) $4x^2 - 4x + 1 = (2x)^2 - 2 \cdot 2x \cdot 1 + 1^2 = (2x - 1)^2$

(4) $x^2 + 10x + 25 = x^2 + 2 \cdot x \cdot 5 + 5^2 = (x + 5)^2$

(5) $x^2 + x + \frac{1}{4} = x^2 + 2 \cdot x \cdot \frac{1}{2} + \left(\frac{1}{2}\right)^2 = \left(x + \frac{1}{2}\right)^2$

(6) $x^6 + 6x^3 + 9 = (x^3)^2 + 2 \cdot x^3 \cdot 3 + 3^2 = (x^3 + 3)^2$

(7) $x^{2n} - 2x^n y^n + y^{2n} = (x^n)^2 - 2 \cdot x^n \cdot y^n + (y^n)^2 = (x^n - y^n)^2$

(8) $x^2 - 1 = x^2 - 1^2 = (x + 1)(x - 1)$

(9) $x^2 - 4y^2 = x^2 - (2y)^2 = (x + 2y)(x - 2y)$

(10) $9x^2 - \frac{1}{4}y^2 = (3x)^2 - \left(\frac{1}{2}y\right)^2 = \left(3x + \frac{1}{2}y\right)\left(3x - \frac{1}{2}y\right)$

(11) $(x + a)^2 - 36 = (x + a)^2 - 6^2 = (x + a + 6)(x + a - 6)$

(12) $1 - (x + y)^2 = 1^2 - (x + y)^2 = (1 + x + y)(1 - x - y)$

(13) $x^4 - 1 = (x^2)^2 - 1^2 = (x^2 + 1)(x^2 - 1) = (x^2 + 1)(x + 1)(x - 1)$

(14) $x^8 - y^8 = (x^4)^2 - (y^4)^2 = (x^4 + y^4)(x^4 - y^4) = (x^4 + y^4)((x^2)^2 - (y^2)^2)$

$$= (x^4 + y^4)(x^2 + y^2)(x^2 - y^2) = (x^4 + y^4)(x^2 + y^2)(x + y)(x - y)$$

(15) $-16x^2 + 9y^2 = (3y)^2 - (4x)^2 = (3y + 4x)(3y - 4x)$

(16) $\frac{1}{9}x^2 - \frac{9}{16}y^2 = \left(\frac{1}{3}x\right)^2 - \left(\frac{3}{4}y\right)^2 = \left(\frac{1}{3}x + \frac{3}{4}y\right)\left(\frac{1}{3}x - \frac{3}{4}y\right)$

(17) $4x^3 y - xy^3 = xy(4x^2 - y^2) = xy(2x + y)(2x - y)$

(18) $x^2 + 4x + 3$

$$\begin{matrix} 1 \\ 1 \end{matrix} \diagdown\diagup \begin{matrix} 1 \to 1 \\ 3 \to 3 \end{matrix} \Big] \xrightarrow{+} 4$$

$= (x + 1)(x + 3)$

(19) $x^2 - 4x - 5$

$$\begin{matrix} 1 \\ 1 \end{matrix} \diagdown\diagup \begin{matrix} 1 \to 1 \\ -5 \to -5 \end{matrix} \Big] \xrightarrow{+} -4$$

$= (x + 1)(x - 5)$

(20) $x^2 - 3x + 2$

$$\begin{matrix} 1 \\ 1 \end{matrix} \diagdown\diagup \begin{matrix} -1 \to -1 \\ -2 \to -2 \end{matrix} \Big] \xrightarrow{+} -3$$

$= (x - 1)(x - 2)$

(21) $x^2 - 2x - 8$

$$\begin{matrix} 1 \\ 1 \end{matrix} \diagdown\diagup \begin{matrix} -4 \to -4 \\ 2 \to 2 \end{matrix} \Big] \xrightarrow{+} -2$$

$= (x - 4)(x + 2)$

(22) $(x - y)^2 - (x + y)^2 = (x - y + x + y)(x - y - x - y) = 2x(-2y) = -4xy$

(23) $(2x - 3)^2 - (x + 1)^2 = (2x - 3 + x + 1)(2x - 3 - x - 1) = (3x - 2)(x - 4)$

(24) $\quad x^2 - \dfrac{5}{6}x + \dfrac{1}{6}$

$$\begin{array}{l} 1 \\ 1 \end{array} \diagdown\!\!\!\diagup \begin{array}{l} -\frac{1}{2} \;\to -\frac{1}{2} \\ -\frac{1}{3} \;\to\; -\frac{1}{3} \end{array} \Bigg] \;\underset{+}{\to}\; -\dfrac{5}{6}$$

$$= \left(x - \dfrac{1}{2}\right)\left(x - \dfrac{1}{3}\right)$$

(25) $\quad 3x^2 - 4x - 4$

$$\begin{array}{l} 1 \\ 3 \end{array} \diagdown\!\!\!\diagup \begin{array}{l} -2 \;\to -6 \\ 2 \;\to\; 2 \end{array} \Bigg] \;\underset{+}{\to}\; -4$$

$$= (x - 2)(3x + 2)$$

(26) $\quad 2x^2 + 3x - 2$

$$\begin{array}{l} 1 \\ 2 \end{array} \diagdown\!\!\!\diagup \begin{array}{l} 2 \;\to 4 \\ -1 \;\to\; -1 \end{array} \Bigg] \;\underset{+}{\to}\; 3$$

$$= (x + 2)(2x - 1)$$

(27) $\quad 4x^2 - 2x - 12$

$$\begin{array}{l} 2 \\ 2 \end{array} \diagdown\!\!\!\diagup \begin{array}{l} 3 \;\to 6 \\ -4 \;\to\; -8 \end{array} \Bigg] \;\underset{+}{\to}\; -2$$

$$= (2x + 3)(2x - 4)$$

(28) $\quad 4x^2 - 10x - 6$

$$= 2(2x^2 - 5x - 3) \quad \text{; factor first}$$

$$\begin{array}{l} 1 \\ 2 \end{array} \diagdown\!\!\!\diagup \begin{array}{l} -3 \;\to -6 \\ 1 \;\to\; 1 \end{array} \Bigg] \;\underset{+}{\to}\; -5$$

$$= 2(x - 3)(2x + 1)$$

(29) $\quad 2x^2 - 3xy - 9y^2$

$$\begin{array}{l} 1 \\ 2 \end{array} \diagdown\!\!\!\diagup \begin{array}{l} -3y \;\to -6y \\ 3y \;\to\; 3y \end{array} \Bigg] \;\underset{+}{\to}\; -3y$$

$$= (x - 3y)(2x + 3y)$$

(30) $\quad x^3y - 16xy^3 = xy(x^2 - 16y^2) = xy(x + 4y)(x - 4y)$

(31) $\quad \dfrac{1}{3}x^2 - 2 + \dfrac{3}{x^2} = \dfrac{1}{3}\left(x^2 - 6 + \dfrac{9}{x^2}\right) = \dfrac{1}{3}\left(x - \dfrac{3}{x}\right)^2$

(32) $\quad 4x^2 - 12xy + 9y^2$

$$\begin{array}{l} 2 \\ 2 \end{array} \diagdown\!\!\!\diagup \begin{array}{l} -3y \;\to -6y \\ -3y \;\to\; -6y \end{array} \Bigg] \;\underset{+}{\to}\; -12y$$

$$= (2x - 3y)(2x - 3y) = (2x - 3y)^2$$

(33) $\quad \dfrac{1}{3}x^2 - \dfrac{1}{3}x - 2$

$$\begin{array}{l} 1 \\ \frac{1}{3} \end{array} \diagdown\!\!\!\diagup \begin{array}{l} 2 \;\to \frac{2}{3} \\ -1 \;\to\; -1 \end{array} \Bigg] \;\underset{+}{\to}\; -\dfrac{1}{3}$$

$$= (x + 2)\left(\dfrac{1}{3}x - 1\right) = \dfrac{1}{3}(x + 2)(x - 3)$$

OR $\quad \dfrac{1}{3}x^2 - \dfrac{1}{3}x - 2 = \dfrac{1}{3}(x^2 - x - 6) = \dfrac{1}{3}(x - 3)(x + 2)$

(34) $a^4(x-y) + b^4(y-x) = (x-y)(a^4 - b^4) = (x-y)(a^2 + b^2)(a^2 - b^2)$

$$= (x-y)(a^2 + b^2)(a+b)(a-b)$$

(35) $-3a^2 + 3a + 6 = -3(a^2 - a - 2) = -3(a-2)(a+1)$ or

$$\begin{array}{c} 1 \\ -3 \end{array} \!\!\!\!\!\!\times\!\!\!\!\!\! \begin{array}{c} -2 \to 6 \\ -3 \to -3 \end{array} \Big] \underset{+}{\to} 3$$

$$= (a-2)(-3a-3) = -3(a-2)(a+1)$$

(36) $3ax^2 - 5ax - 2a$

$$= a(3x^2 - 5x - 2)$$

$$\begin{array}{c} 1 \\ 3 \end{array} \!\!\!\!\!\!\times\!\!\!\!\!\! \begin{array}{c} -2 \to -6 \\ 1 \to 1 \end{array} \Big] \underset{+}{\to} -5$$

$$= a(x-2)(3x+1)$$

(37) $(a-1)^2 - 10(a-1) + 25$

$$\begin{array}{c} a-1 \\ a-1 \end{array} \!\!\!\!\!\!\times\!\!\!\!\!\! \begin{array}{c} -5 \to -5(a-1) \\ -5 \to -5(a-1) \end{array} \Big] \underset{+}{\to} -10(a-1)$$

$$= (a-1-5)(a-1-5) = (a-6)^2$$

OR let $x = a - 1$. Then

$$(a-1)^2 - 10(a-1) + 25 = x^2 - 10x + 25 = (x-5)^2 = (a-1-5)^2 = (a-6)^2$$

(38) $a^8 - 1 = (a^4)^2 - 1^2 = (a^4 + 1)(a^4 - 1) = (a^4 + 1)(a^2 + 1)(a^2 - 1)$

$$= (a^4 + 1)(a^2 + 1)(a+1)(a-1)$$

(39) $(x-1)(x+2) - 1 - x^2 \mid x \quad 2 \quad 1 = x^2 \mid x - 6 - (x+3)(x-2)$

(40) $4x^2 - 2x - 2 - (x-1)^2 = 4x^2 - 2x - 2 - (x^2 - 2x + 1) = 3x^2 - 3$

$$= 3(x^2 - 1) = 3(x+1)(x-1)$$

#4. Find the value of k that will make the following polynomials perfect square forms.

(1) $x^2 + 5x + k = x^2 + 2 \cdot x \cdot \frac{5}{2} + \left(\frac{5}{2}\right)^2 = \left(x + \frac{5}{2}\right)^2 \therefore k = \frac{25}{4}$

OR $x^2 + 5x + k = (x + \frac{5}{2})^2 = x^2 + 5x + \left(\frac{5}{2}\right)^2 \therefore k = \frac{25}{4}$

(2) $9x^2 - 12x + k = (3x)^2 - 2 \cdot 3x \cdot 2 + (2)^2 = (3x - 2)^2 \therefore k = 4$

OR $9x^2 - 12x + k = 9\left(x^2 - \frac{12}{9}x + \frac{k}{9}\right) = 9\left(x^2 - \frac{4}{3}x + \frac{k}{9}\right) = 9\left(x - \frac{1}{2} \cdot \frac{4}{3}\right)^2 = 9\left(x - \frac{2}{3}\right)^2$

$$= 9\left(x^2 - \frac{4}{3}x + \frac{4}{9}\right) = 9x^2 - 12x + 4 \therefore k = 4$$

(3) $2x^2 - 6x + k = 2\left(x^2 - 3x + \frac{k}{2}\right) = 2\left(x - \frac{1}{2} \cdot 3\right)^2 = 2\left(x - \frac{3}{2}\right)^2 = 2\left(x^2 - 3x + \frac{9}{4}\right)$

$$= 2x^2 - 6x + \frac{9}{2} \quad \therefore k = \frac{9}{2}$$

(4) $25x^2 + 4x + k = (5x)^2 + 2 \cdot 5x \cdot \frac{2}{5} + \left(\frac{2}{5}\right)^2 = \left(5x + \frac{2}{5}\right)^2 \quad \therefore k = \frac{4}{25}$

OR $25x^2 + 4x + k = 25\left(x^2 + \frac{4}{25}x + \frac{k}{25}\right) = 25\left(x + \frac{1}{2} \cdot \frac{4}{25}\right)^2 = 25\left(x + \frac{2}{25}\right)^2$

$$= 25\left(x^2 + \frac{4}{25}x + \left(\frac{2}{25}\right)^2\right) = 25x^2 + 4x + \frac{4}{25} \quad \therefore k = \frac{4}{25}$$

(5) $\frac{1}{25}x^2 + kx + 4 = \left(\frac{1}{5}x \pm 2\right)^2 = \frac{1}{25}x^2 \pm \frac{4}{5}x + 4 \quad \therefore k = \frac{4}{5} \ \text{ or } \ k = -\frac{4}{5}$

(6) $4x^2 - kx + 25 = (2x \pm 5)^2 = 4x^2 \pm 20x + 25 \quad \therefore -k = \pm 20 \ ; \ k = -20 \ \text{ or } \ k = 20$

(7) $2x^2 + kx + 8 = 2\left(x^2 + \frac{k}{2}x + 4\right) = 2\left(x + \frac{k}{4}\right)^2 = 2\left(x^2 + \frac{k}{2}x + \frac{k^2}{16}\right) = 2x^2 + kx + \frac{k^2}{8}$

$\therefore \frac{k^2}{8} = 8 \ ; \ k^2 = 64 \ ; \quad k = 8 \ \text{ or } \ k = -8$

OR $2\left(x^2 + \frac{k}{2}x + 4\right) = 2(x \pm 2)^2 = 2(x^2 \pm 4x + 4) \quad \therefore \frac{k}{2} = \pm 4 \ ; \ k = \pm 8$

(8) $9x^2 + (2k - 4)x + 4 = (3x \pm 2)^2 = 9x^2 \pm 12x + 4 \quad \therefore 2k - 4 = \pm 12 \ ; \ 2k = 16 \ \text{ or } \ 2k = -8$

$\therefore \ k = 8 \ \text{ or } \ k = -4$

(9) $4x^2 + (k + 5)x + 9y^2 = (2x \pm 3y)^2 = 4x^2 \pm 12xy + 9y^2$

$\therefore k + 5 = \pm 12y \ ; \ k = 12y - 5 \ \text{or } \ k = -12y - 5$

(10) $k - \frac{1}{4}xy + \frac{1}{4}y^2 = \frac{1}{4}y^2 - \frac{1}{4}xy + k = \left(\frac{1}{2}y\right)^2 - 2\left(\frac{1}{2}y\right)\left(\frac{1}{4}x\right) + \left(\frac{1}{4}x\right)^2 = \left(\frac{1}{2}y - \frac{1}{4}x\right)^2$

$$= \frac{1}{4}y^2 - \frac{1}{4}xy + \frac{1}{16}x^2 \quad \therefore k = \frac{1}{16}x^2$$

(11) $9x^2 + (k - 1)xy + 25y^2 = (3x \pm 5y)^2 = 9x^2 \pm 30xy + 25y^2$

$\therefore k - 1 = \pm 30 \ ; \ k = 31 \ \text{or } k = -29$

(12) $kx^2 + 3x + 9 = k\left(x^2 + \frac{3}{k}x + \frac{9}{k}\right) = k\left(x + \frac{3}{2k}\right)^2 = k\left(x^2 + \frac{3}{k}x + \frac{9}{4k^2}\right) = kx^2 + 3x + \frac{9}{4k}$

$\therefore \frac{9}{4k} = 9 \ ; \ 4k = 1 \ ; \ k = \frac{1}{4}$

(13) $(x + 3)(x - 4) - k = x^2 - x - 12 - k = x^2 - 2 \cdot x \cdot \frac{1}{2} + \left(\frac{1}{2}\right)^2 = \left(x - \frac{1}{2}\right)^2$

$\therefore \frac{1}{4} = -12 - k \ ; \ k = -12 - \frac{1}{4} = -12\frac{1}{4}$

(14) $(2x + 1)(2x - 4) + k = 4x^2 + 2x - 8x - 4 + k = 4x^2 - 6x - 4 + k$

$$= (2x)^2 - 2 \cdot 2x \cdot \frac{6}{4} + \left(\frac{6}{4}\right)^2 = \left(2x - \frac{6}{4}\right)^2$$

$$\therefore \left(\frac{6}{4}\right)^2 = -4 + k \; ; k = \frac{9}{4} + 4 = 6\frac{1}{4}$$

(15) $(x - 1)(x - 2)(x + 4)(x + 5) + k = (x - 1)(x + 4)(x - 2)(x + 5) + k$

$$= (x^2 + 3x - 4)(x^2 + 3x - 10) + k = (A - 4)(A - 10) + k, \quad \text{letting } x^2 + 3x = A$$

$$= A^2 - 14A + 40 + k = (A - 7)^2 = A^2 - 14A + 49 \quad \therefore \quad 40 + k = 49 \; ; \; k = 9$$

#5. Find the value of a for the following polynomials. Each polynomial has a given factor.

(1) $x^2 + 2x + a$ has the factor $(x + 3)$.

Let $x^2 + 2x + a = (x + 3)(x + b)$

$\Rightarrow x^2 + 2x + a = x^2 + (3 + b)x + 3b$

Since $3 + b = 2$ and $3b = a$, $b = -1$ and $a = -3$ $\therefore a = -3$

(OR Since $x + 3$ is a factor of $x^2 + 2x + a$, $(-3)^2 + 2(-3) + a = 0$.

So, $9 - 6 + a = 0$ $\therefore a = -3$)

(2) $3x^2 + ax - 8$ has the factor $(x - 2)$.

Let $3x^2 + ax - 8 = (x - 2)(3x + b)$

$\Rightarrow 3x^2 + ax - 8 = 3x^2 + (b - 6)x - 2b$

Since $b - 6 = a$ and $-2b = -8$, $b = 4$ and $a = -2$ $\therefore a = -2$

(OR Since $x - 2$ is a factor of $3x^2 + ax - 8$, $3(2)^2 + 2a - 8 = 0$.

So, $4 + 2a = 0$ $\therefore a = -2$)

(3) $4x^2 + ax - 6$ has the factor $(3 - 2x)$.

Let $4x^2 + ax - 6 = (3 - 2x)(b - 2x)$

$\Rightarrow 4x^2 + ax - 6 = 4x^2 + (-2b - 6)x + 3b$

Since $-2b - 6 = a$ and $3b = -6$, $b = -2$ and $a = -2$ $\therefore a = -2$

(OR Since $3 - 2x$ is a factor of $4x^2 + ax - 6$, $4\left(\frac{3}{2}\right)^2 + a\left(\frac{3}{2}\right) - 6 = 0$.

So, $3 + \frac{3}{2}a = 0$ $\therefore a = -2$)

(4) $2x^2 + (3a - 1)x - 15$ **has the factor** $(2x + 3)$.

Let $2x^2 + (3a - 1)x - 15 = (2x + 3)(x + b)$

$\Rightarrow 2x^2 + (3a - 1)x - 15 = 2x^2 + (3 + 2b)x + 3b$

Since $3 + 2b = 3a - 1$ and $3b = -15$, $b = -5$ and $a = -2$ $\therefore a = -2$

(OR Since $2x + 3$ is a factor of $2x^2 + (3a - 1)x - 15$, $2\left(-\dfrac{3}{2}\right)^2 + (3a - 1)\left(-\dfrac{3}{2}\right) - 15 = 0$.

So, $\dfrac{9}{2} - \dfrac{9}{2}a + \dfrac{3}{2} - 15 = 0$; $9 - 9a + 3 - 30 = 0$; $9a = -18$ $\therefore a = -2$)

(5) $2ax^2 - 5x + 2$ **has the factor** $(3x - 2)$.

Let $2ax^2 - 5x + 2 = (3x - 2)(bx - 1)$

$\Rightarrow 2ax^2 - 5x + 2 = 3bx^2 + (-2b - 3)x + 2$

Since $3b = 2a$ and $-2b - 3 = -5$, $b = 1$ and $a = \dfrac{3}{2}$ $\therefore a = \dfrac{3}{2}$

(OR Since $3x - 2$ is a factor of $2ax^2 - 5x + 2$, $2a\left(\dfrac{2}{3}\right)^2 - 5\left(\dfrac{2}{3}\right) + 2 = 0$.

So, $\dfrac{8a}{9} - \dfrac{10}{3} + 2 = 0$; $8a - 30 + 18 = 0$; $8a = 12$ $\therefore a = \dfrac{3}{2}$)

#6. Find the value of $a + b$ for any constants a and b.

(1) $3ax^2 - 6x + ab$ **has the factor** $(3x - 1)^2$.

$3ax^2 - 6x + ab = c(3x - 1)^2 = 9cx^2 - 6cx + c$ $\therefore 3a = 9c, -6 = -6c$, and $ab = c$

$\therefore c = 1, a = 3, b = \dfrac{1}{3}$ Therefore, $a + b = 3\dfrac{1}{3}$

(2) $ax^2 + 8x + 4b$ **has two factors** $(3x + 2)$ **and** $(x - 2)$.

$ax^2 + 8x + 4b = c(3x + 2)(x - 2) = 3cx^2 - 4cx - 4c$ $\therefore a = 3c, 8 = -4c$, and $4b = -4c$

$\therefore c = -2, b = 2, a = -6$ Therefore, $a + b = -4$

(3) $(4x - 3)^2 - (3x - 2)^2$ **has two factors** $(ax + 5)$ **and** $(b - x)$.

$(4x - 3)^2 - (3x - 2)^2 = (4x - 3 + 3x - 2)(4x - 3 - 3x + 2) = (7x - 5)(x - 1)$

$= -(-7x + 5)(x - 1) = (-7x + 5)(1 - x) = (ax + 5)(b - x)$

$\therefore a = -7$ and $b = 1$

Therefore, $a + b = -6$

(4) $2x^2 + ax - 4$ and $bx^2 - x - 2$ have the same factor $(2x + 1)$.

$2x^2 + ax - 4 = (2x + 1)(x + c) = 2x^2 + (1 + 2c)x + c$ $\therefore a = 1 + 2c$ and $-4 = c$ $\therefore a = -7$

$bx^2 - x - 2 = (2x + 1)(dx - 2) = 2dx^2 + (d - 4)x - 2$ $\therefore b = 2d$ and $-1 = d - 4$

$\therefore d = 3$ and $b = 6$

Therefore, $a + b = -1$

(OR $2x + 1 = 0$; $x = -\frac{1}{2}$

$2x^2 + ax - 4 = 2\left(-\frac{1}{2}\right)^2 + a\left(-\frac{1}{2}\right) - 4 = 0$; $\frac{1}{2} - \frac{1}{2}a - 4 = 0$ $\therefore a = -7$

$bx^2 - x - 2 = b\left(-\frac{1}{2}\right)^2 - \left(-\frac{1}{2}\right) - 2 = 0$; $\frac{1}{4}b + \frac{1}{2} - 2 = 0$ $\therefore b = 6$)

#7. For any integers $a(\neq 0)$ and b, the length and width of a rectangle are forms of $ax + b$.

Find the perimeter of a rectangle whose area is $6x^2 + 17x + 12$.

$6x^2 + 17x + 12$

$\begin{array}{l} 2 \quad\quad\quad 3 \to 9 \\ 3 \quad\quad\quad 4 \to 8 \end{array}] \xrightarrow[+]{} 17$

$= (2x + 3)(3x + 4)$

$\therefore$ The perimeter is $2(2x + 3) + 2(3x + 4) = 10x + 14$.

#8. Find a possible expression for the width of a rectangle whose area is $12x^2 + 5x - a$, where $a > 0$ and the width is greater than the length.

$12x^2 + 5x - a = (3x - 1)(4x + a) = 12x^2 - 4x + 3ax - a = 12x^2 + (-4 + 3a)x - a$

$\therefore -4 + 3a = 5$; $3a = 9$; $a = 3$

$\therefore$ The width is $4x + 3$.

#9. **The area and base of a triangle are $5x^2 + 12x + 4$ and $2x + 4$, respectively.**

Find the height of the triangle.

$5x^2 + 12x + 4 = \frac{1}{2} \cdot (2x + 4) \cdot \text{height} = (x + 2) \cdot \text{height}$

$5x^2 + 12x + 4$

$= (x + 2)(5x + 2)$

∴ The height is $5x + 2$.

#10. **The figures $A, B,$ and C(equilateral) have the same area. Find the perimeters of A and C.**

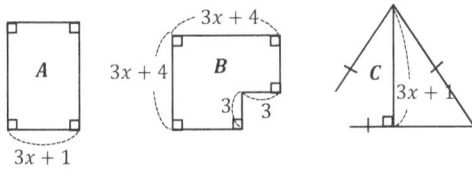

The area of $B = (3x + 4)^2 - 9 = (3x + 4 + 3)(3x + 4 - 3) = (3x + 7)(3x + 1)$

$=$ The area of $A =$ The area of C ∴ The width of A is $3x + 7$.

Therefore, the perimeter of A is $2(3x + 1) + 2(3x + 7) = 12x + 16$.

Since the area of C is $\frac{1}{2} \cdot \text{base} \cdot (3x + 1) = (3x + 7)(3x + 1)$, the base of C is $2(3x + 7)$.

Therefore, the perimeter of C is $3 \cdot 2(3x + 7) = 18x + 42$.

#11. **A and B are squares with the lengths a and b, respectively. The perimeter of A is 40 more than the perimeter of B. The difference between their areas is 200. Find the sum of their areas.**

$4a = 4b + 40$; $4(a - b) = 40$; $a - b = 10$

$a^2 - b^2 = 200$; $(a + b)(a - b) = 200$ ∴ $a + b = 20$

$\begin{cases} a + b = 20 \\ a - b = 10 \end{cases} \Rightarrow 2a = 30$; $a = 15$ ∴ $b = 5$

Therefore, the sum of their areas is $15^2 + 5^2 = 250$.

(OR the sum of their areas is $a^2 + b^2 = (a + b)^2 - 2ab = 20^2 - 2 \cdot 15 \cdot 5 = 400 - 150 = 250$).

#12. Factor each polynomial using any method.

(1) $3a^2b - a^3b - 2ab = ab(3a - a^2 - 2) = -ab(a^2 - 3a + 2) = -ab(a - 2)(a - 1)$

(2) $3a - 3a^2 + 6 = -3(a^2 - a - 2) = -3(a - 2)(a + 1)$

(3) $8a^2b + 6ab - 20b = 2b(4a^2 + 3a - 10) = 2b(a + 2)(4a - 5)$

(4) $a^5 - 16a = a(a^4 - 16) = a(a^2 + 4)(a^2 - 4) = a(a^2 + 4)(a + 2)(a - 2)$

(5) $a^2(x - y) + b^2(y - x) = a^2(x - y) - b^2(x - y) = (x - y)(a^2 - b^2) = (x - y)(a + b)(a - b)$

(6) $ab(a - b) + 2a(b - a)^2 = ab(a - b) + 2a(a - b)^2$ $\quad (\because (a - b)^2 = \left(-(a - b)\right)^2 = (b - a)^2)$

$\qquad = (a - b)\bigl(ab + 2a(a - b)\bigr)$

$\qquad = a(a - b)(b + 2a - 2b) = a(a - b)(2a - b)$

(7) $(a - b)^2 - 2(b - a)^3 = (a - b)^2 + 2(a - b)^3$ $\quad (\because (a - b)^3 = \left(-(b - a)\right)^3 = -(b - a)^3)$

$\qquad = (a - b)^2\bigl(1 + 2(a - b)\bigr) = (a - b)^2(1 + 2a - 2b)$

(8) $a(x - y)^2 + b(x - y) = (x - y)(a(x - y) + b) = (x - y)(ax - ay + b)$

(9) $4(x + 2)^2 - 3(x + 2) - 10 = 4A^2 - 3A - 10$, letting $x + 2 = A$,

$\quad 4A^2 - 3A - 10$

$\quad \begin{matrix} 1 \\ 4 \end{matrix} \diagdown\!\!\!\diagup \begin{matrix} -2 \;\to\; -8 \\ 5 \;\to\; 5 \end{matrix} \Big] \underset{+}{\to} -3$

$\quad = (A - 2)(4A + 5)$

$\quad = (x + 2 - 2)(4(x + 2) + 5) = x(4x + 13)$

(10) $(x - y)(x - y + 3) - 4 = A(A + 3) - 4$, letting $x - y = A$,

$\quad = A^2 + 3A - 4$

$\quad \begin{matrix} 1 \\ 1 \end{matrix} \diagdown\!\!\!\diagup \begin{matrix} -1 \;\to\; -1 \\ 4 \;\to\; 4 \end{matrix} \Big] \underset{+}{\to} 3$

$\quad = (A - 1)(A + 4)$

$\quad = (x - y - 1)(x - y + 4)$

(11) $(2x - y)^2 + 8y(2x - y) + 16y^2 = A^2 + 8yA + 16y^2$, letting $2x - y = A$,

$\quad A^2 + 8yA + 16y^2$

$\quad \begin{matrix} 1 \\ 1 \end{matrix} \diagdown\!\!\!\diagup \begin{matrix} 4y \;\to\; 4y \\ 4y \;\to\; 4y \end{matrix} \Big] \underset{+}{\to} 8y$

$\quad = (A + 4y)(A + 4y) = (A + 4y)^2 = (2x - y + 4y)^2 = (2x + 3y)^2$

(12) $2(x - 1)^2 - 3(x - 1)(y + 1) - 9(y + 1)^2 = 2A^2 - 3AB - 9B^2$, letting $x - 1 = A,\ y + 1 = B$,

$\quad 2A^2 - 3AB - 9B^2$

$\quad \begin{matrix} 1 \\ 2 \end{matrix} \diagdown\!\!\!\diagup \begin{matrix} -3B \;\to\; -6B \\ 3B \;\to\; 3B \end{matrix} \Big] \underset{+}{\to} -3B$

$\quad = (A - 3B)(2A + 3B) = \bigl(x - 1 - 3(y + 1)\bigr)\bigl(2(x - 1) + 3(y + 1)\bigr) = (x - 3y - 4)(2x + 3y + 1)$

(13) $a^4 - 5a^2 - 36 = (a^2)^2 - 5a^2 - 36 = A^2 - 5A - 36$, letting $A = a^2$,

$\quad = (A - 9)(A + 4) = (a^2 - 9)(a^2 + 4)$

$\quad = (a + 3)(a - 3)(a^2 + 4)$

(14) $a^8 - 2a^4 - 8 = (a^4)^2 - 2a^4 - 8 = A^2 - 2A - 8$, letting $A = a^4$,

$\quad = (A - 4)(A + 2) = (a^4 - 4)(a^4 + 2)$

$\quad = (a^2 + 2)(a^2 - 2)(a^4 + 2)$

(15) $x^2 - xy + 2x - 2y = (x^2 - xy) + 2(x - y) = x(x - y) + 2(x - y) = (x - y)(x + 2)$

(16) $a^2 - ab - b - 1 = -b(a + 1) + a^2 - 1 = -b(a + 1) + (a + 1)(a - 1)$

$\quad = (a + 1)(-b + a - 1) = (a + 1)(a - b - 1)$

(17) $a^3 - a^2 - a + 1 = a^2(a - 1) - (a - 1) = (a - 1)(a^2 - 1) = (a - 1)(a + 1)(a - 1)$

$\quad = (a - 1)^2(a + 1)$

(18) $a^4 + 3a - 3a^3 - a^2 = a^4 - 3a^3 - a^2 + 3a = a^3(a - 3) - a(a - 3) = a(a - 3)(a^2 - 1)$

$\quad = a(a - 3)(a + 1)(a - 1)$

(19) $2ab + 1 - a^2 - b^2 = 1 - (a^2 + b^2 - 2ab) = 1 - (a - b)^2 = (1 + a - b)(1 - a + b)$

(20) $(x + 2)(x - 1)^2(x - 4) - 10 = (x - 1)^2(x + 2)(x - 4) - 10$

$\quad = (x^2 - 2x + 1)(x^2 - 2x - 8) - 10 = (A + 1)(A - 8) - 10$, letting $A = x^2 - 2x$,

$\quad = A^2 - 7A - 8 - 10 = A^2 - 7A - 18 = (A - 9)(A + 2)$

$\quad = (x^2 - 2x - 9)(x^2 - 2x + 2)$

(21) $9x^2 - y^2 - 4y - 4 = 9x^2 - (y^2 + 4y + 4) = (3x)^2 - (y + 2)^2 = (3x + y + 2)(3x - y - 2)$

(22) $a^2 - b^2 + 4b - 4 = a^2 - (b^2 - 4b + 4) = a^2 - (b - 2)^2 = (a + b - 2)(a - b + 2)$

(23) $a^2 - 16b^2 - 6a + 9 = (a^2 - 6a + 9) - 16b^2 = (a - 3)^2 - (4b)^2$

$\quad = (a - 3 + 4b)(a - 3 - 4b) = (a + 4b - 3)(a - 4b - 3)$

(24) $ax^2 - a - bx^2 + b = (a - b)x^2 - (a - b) = (a - b)(x^2 - 1) = (a - b)(x + 1)(x - 1)$

(25) $x^2 - xy + x + 2y - 6 = y(-x + 2) + x^2 + x - 6 = -y(x - 2) + (x + 3)(x - 2)$

$\quad = (x - 2)(-y + x + 3) = (x - 2)(x - y + 3)$

(26) $-2a^2 - 5a + 2ab - b + 3 = b(2a - 1) - (2a^2 + 5a - 3) = b(2a - 1) - (a + 3)(2a - 1)$

$\quad = (2a - 1)(b - a - 3)$

(27) $2x^2 - y^2 + xy - 2x + y = 2x^2 + (y-2)x - y^2 + y = 2x^2 + (y-2)x - y(y-1)$

$= (x + y - 1)(2x - y)$

(28) $4a^2 - b^2 - 4a + 4b - 3 = -(b^2 - 4b + 4) + 4a^2 - 4a - 3 + 4$

$= -(b-2)^2 + 4a^2 - 4a + 1 = -(b-2)^2 + (2a-1)^2 = (2a-1)^2 - (b-2)^2$

$= (2a - 1 + b - 2)(2a - 1 - b + 2) = (2a + b - 3)(2a - b + 1)$

(29) $4a^2 - 4ab + b^2 - c^2 = (4a^2 - 4ab + b^2) - c^2 = (2a-b)^2 - c^2 = (2a - b + c)(2a - b - c)$

(30) $a^4 + a^2 + 1 = A^2 + A + 1$, letting $a^2 = A$,

$= A^2 + 2A + 1 - A = (A+1)^2 - A = (a^2 + 1)^2 - a^2 = (a^2 + 1 + a)(a^2 + 1 - a)$

$= (a^2 + a + 1)(a^2 - a + 1)$

(31) $a^4 - 6a^2 + 1 = (A^2 - 2A + 1) - 4A$, letting $a^2 = A$,

$= (A-1)^2 - 4A = (a^2 - 1)^2 - 4a^2 = (a^2 - 1 + 2a)(a^2 - 1 - 2a) = (a^2 + 2a - 1)(a^2 - 2a - 1)$

(32) $a^4 - 13a^2 + 4 = (A^2 - 4A + 4) - 9A$, letting $a^2 = A$,

$= (A-2)^2 - 9A = (a^2 - 2)^2 - (3a)^2 = (a^2 - 2 + 3a)(a^2 - 2 - 3a)$

$= (a^2 + 3a - 2)(a^2 - 3a - 2)$

(33) $9x^4 + 8x^2 + 4 = 9A^2 + 8A + 4$, letting $x^2 = A$,

$= 9A^2 + 12A + 4 - 4A = (3A + 2)^2 - 4A = (3x^2 + 2)^2 - (2x)^2 = (3x^2 + 2 + 2x)(3x^2 + 2 - 2x)$

$= (3x^2 + 2x + 2)(3x^2 - 2x + 2)$

#13. **Evaluate each expression using factorization.**

(1) $99^2 - 1 = (99 + 1)(99 - 1) = 100 \cdot 98 = 9800$

(2) $99^2 - 89^2 = (99 + 89)(99 - 89) = 188 \cdot 10 = 1880$

(3) $49^2 - 51^2 = (49 + 51)(49 - 51) = 100 \cdot -2 = -200$

(4) $3^8 - 1 = (3^4)^2 - 1 = (3^4 + 1)(3^4 - 1) = (3^4 + 1)(3^2 + 1)(3^2 - 1)$

$= (3^4 + 1)(3^2 + 1)(3 + 1)(3 - 1) = 82 \cdot 10 \cdot 4 \cdot 2 = 6560$

(5) $6^2 - 5^2 + 4^2 - 3^2 + 2^2 - 1 = (6 + 5)(6 - 5) + (4 + 3)(4 - 3) + (2 + 1)(2 - 1)$

$= 11 \cdot 1 + 7 \cdot 1 + 3 \cdot 1 = 21$

(6) $\left(1 - \frac{1}{2^2}\right)\left(1 - \frac{1}{3^2}\right)\left(1 - \frac{1}{4^2}\right)\cdots\cdots\left(1 - \frac{1}{99^2}\right)\left(1 - \frac{1}{100^2}\right)$

$= \left(1 - \frac{1}{2}\right)\left(1 + \frac{1}{2}\right)\left(1 - \frac{1}{3}\right)\left(1 + \frac{1}{3}\right)\left(1 - \frac{1}{4}\right)\left(1 + \frac{1}{4}\right)\cdots\cdots\left(1 - \frac{1}{99}\right)\left(1 + \frac{1}{99}\right)\left(1 - \frac{1}{100}\right)\left(1 + \frac{1}{100}\right)$

$= \left(\frac{1}{2}\cdot\frac{3}{2}\right)\left(\frac{2}{3}\cdot\frac{4}{3}\right)\left(\frac{3}{4}\cdot\frac{5}{4}\right)\cdots\cdots\left(\frac{98}{99}\cdot\frac{100}{99}\right)\left(\frac{99}{100}\cdot\frac{101}{100}\right) = \frac{1}{2}\cdot\frac{101}{100} = \frac{101}{200}$

(7) $3(2^2 + 1)(2^4 + 1)(2^8 + 1) + 1 = (2^2 - 1)(2^2 + 1)(2^4 + 1)(2^8 + 1) + 1$

$= (2^4 - 1)(2^4 + 1)(2^8 + 1) + 1 = (2^8 - 1)(2^8 + 1) + 1 = (2^{16} - 1) + 1 = 2^{16}$

(8) $\frac{99 \times 101 + 99 \times 2}{101^2 - 4} = \frac{99\,(101 + 2)}{(101 + 2)(101 - 2)} = \frac{99}{(101 - 2)} = \frac{99}{99} = 1$

(9) $36 \times 34 - 35 \times 34 = 34(36 - 35) = 34 \cdot 1 = 34$

(10) $87 \times 56 + 87 \times 44 = 87(56 + 44) = 87 \cdot 100 = 8700$

(11) $65^2 - 2 \times 65 \times 35 + 35^2 = (65 - 35)^2 = 30^2 = 900$

(12) $25^2 + 30 \times 25 + 15^2 = 25^2 + 2 \cdot 25 \cdot 15 + 15^2 = (25 + 15)^2 = 40^2 = 1600$

(13) $a^2 - 8a - 20$, when $a = 28$

$a^2 - 8a - 20 = (a - 10)(a + 2) = 18 \cdot 30 = 540$

(14) $a^2 + 3a - 54$, when $a = 91$

$a^2 + 3a - 54 = (a + 9)(a - 6) = 100 \cdot 85 = 8500$

Solutions for Chapter 7

#1. State whether each expression is a quadratic equation (Yes) or is not a quadratic equation(No).

(1) $x^2 = 5$; Yes

(2) $x(x + 3) = 0$; Yes

(3) $x^2 = (x + 3)^2$; No

(4) $x^3 + x^2 + 1 = 0$; No

(5) $2x^3 + x^2 = 3x + 2\,x^3 + 5$; Yes

(6) $\frac{1}{x^2} + \frac{1}{x} + 3 = 0$; No

(7) $(x + 1)^2 - (x - 1)^2 + 3 = 0$; No

(8) $2x^2 + 5x = (x + 1)(x + 2)$; Yes

(9) $x^2 + 4x + 4 = 2(x + 2)^2 - 2$; Yes

(10) $x^2 - 5x + 6$; No

(11) $3x^2 + 5x + 1 = 2x^2 - 6x + 4$; Yes

(12) $\frac{x^2}{2} + 4x = 4x + 4$; Yes

(13) $\frac{2}{x^2} + 2x = 3$; No

(14) $x^2 + 1 = (x + 1)^2$; No

(15) $x^2 + 2x + 1 = 2(x + 1)^2$; Yes

(16) $x^2 = 0$; Yes

(17) $x(2x + 1) = 3x(x + 2)$; Yes

(18) $(x + 4)^2$; No

(19) $x^2(x - 1) = x(x^2 + x - 1)$; Yes

(20) $x^2 + 3x = x^2 + 3$; No

#2. The following equations are quadratic equations. Find the condition for constants a and b.

(1) $(x + 1)(ax + 2) = 2x^2 + 5$; $ax^2 \neq 2x^2$; $a \neq 2$

(2) $2(x^2 + 2x + 1) = (x + 2)(5 - ax)$; $2x^2 \neq -ax^2$; $a \neq -2$

(3)　$(3x - 1)(ax + 2) = 5 - bx^2$; $3ax^2 \neq -bx^2$; $(3a + b)x^2 \neq 0$; $3a + b \neq 0$

(4)　$(2x + 1)(3x + 2) = (ax + 2)(bx - 3)$; $6x^2 \neq abx^2$; $ab \neq 6$

(5)　$(2a + b)x^2 + ax + b = 0$; $2a + b \neq 0$

(6)　$a^2x^2 + bx + 5 = 5$; $a^2 \neq 0$; $a \neq 0$

#3.　**Find the value of $a + b + c$ for the quadratic equation $ax^2 + bx + c = 0$, in which a is the smallest positive number.**

(1)　$2(x - 1)^2 = (x + 1)^2 + 5$

　　$2(x^2 - 2x + 1) = x^2 + 2x + 1 + 5$; $x^2 - 6x - 4 = 0$　$\therefore a + b + c = 1 - 6 - 4 = -9$

(2)　$(3x + 1)(x - 2) = 2 - x^2$

　　$3x^2 - 5x - 2 = 2 - x^2$; $4x^2 - 5x - 4 = 0$　$\therefore a + b + c = 4 - 5 - 4 = -5$

(3)　$3(x + 1)^2 = 3(x + 1)$

　　$3x(x + 1) = 0$; $3x^2 + 3x = 0$　$\therefore a + b + c = 3 + 3 + 0 = 6$

(4)　$x^2 = x$

　　$x^2 - x = 0$　$\therefore a + b + c = 1 - 1 + 0 = 0$

(5)　$2x(x - 1) = x^2 - 2$

　　$2x^2 - 2x = x^2 - 2$; $x^2 - 2x + 2 = 0$　$\therefore a + b + c = 1 - 2 + 2 = 1$

#4.　**Find the sum of the solutions for each quadratic equation using factorization.**

(1)　$x^2 - 2x - 3 = 0$; $(x - 3)(x + 1) = 0$; $x = 3, -1$　$\therefore 2$

(2)　$2x^2 - 7x + 5 = 0$; $(x - 1)(2x - 5) = 0$; $x = 1, \frac{5}{2}$　$\therefore 3\frac{1}{2}$

(3)　$-3x^2 + 6x = 0$; $-3x(x - 2) = 0$; $x = 0, 2$　$\therefore 2$

(4)　$2x^2 + 2x - 4 = 0$; $2(x^2 + x - 2) = 0$; $2(x + 2)(x - 1) = 0$; $x = -2, 1$　$\therefore -1$

(5)　$x(x + 5) = 6$; $(x + 6)(x - 1) = 0$; $x = -6, 1$　$\therefore -5$

(6)　$x^2 = \frac{x+1}{2}$; $2x^2 - x - 1 = 0$; $(x - 1)(2x + 1) = 0$; $x = 1, -\frac{1}{2}$　$\therefore \frac{1}{2}$

#5. Each of the following quadratic equations has a solution $x = \alpha$. For each equation, find the value of constant a and the other solution $x = \beta$ for the equation.

    (1) $x^2 + ax - 4 = 0$, $x = 1$

        Substitute $x = 1$ into the equation ; $1 + a - 4 = 0$ $\therefore a = 3$

        Then $x^2 + 3x - 4 = 0$; $(x + 4)(x - 1) = 0$ $\therefore \beta = -4$

    (2) $3x^2 - ax + a = 0$, $x = -1$

        Substitute $x = -1$ into the equation ; $3 + a + a = 0$ $\therefore a = -\dfrac{3}{2}$

        Then $3x^2 + \dfrac{3}{2}x - \dfrac{3}{2} = 0$; $6x^2 + 3x - 3 = 0$; $3(2x^2 + x - 1) = 0$

        $3(2x - 1)(x + 1) = 0$ $\therefore \beta = \dfrac{1}{2}$

    (3) $2x^2 - x + a = 0$, $x = -1$

        Substitute $x = -1$ into the equation ; $2 + 1 + a = 0$ $\therefore a = -3$

        Then $2x^2 - x - 3 = 0$; $(2x - 3)(x + 1) = 0$ $\therefore \beta = \dfrac{3}{2}$

    (4) $ax^2 - 2x - 3 = 0$, $x = -2$

        Substitute $x = -2$ into the equation ; $4a + 4 - 3 = 0$ $\therefore a = -\dfrac{1}{4}$

        Then $-\dfrac{1}{4}x^2 - 2x - 3 = 0$; $x^2 + 8x + 12 = 0$; $(x + 6)(x + 2) = 0$ $\therefore \beta = -6$

    (5) $ax^2 - (a - 1)x - 6 = 0$, $x = 2$

        Substitute $x = 2$ into the equation ; $4a - 2(a - 1) - 6 = 0$; $2a - 4 = 0$ $\therefore a = 2$

        Then $2x^2 - x - 6 = 0$; $(x - 2)(2x + 3) = 0$ $\therefore \beta = -\dfrac{3}{2}$

#6. Each of the following quadratic equations has the solution $x = \alpha$. Find the value of the given expression for each equation.

    (1) $\alpha - \dfrac{1}{\alpha}$ for $x^2 - 2x + 1 = 0$

        $\alpha^2 - 2\alpha + 1 = 0 \Rightarrow \alpha - 2 + \dfrac{1}{\alpha} = 0 \Rightarrow \alpha + \dfrac{1}{\alpha} = 2$

        $\therefore \left(\alpha - \dfrac{1}{\alpha}\right)^2 = \left(\alpha + \dfrac{1}{\alpha}\right)^2 - 4 = 4 - 4 = 0$ $\therefore \alpha - \dfrac{1}{\alpha} = 0$

    (2) $\alpha^2 + \dfrac{1}{\alpha^2}$ for $2x^2 + 3x - 2 = 0$

        $2\alpha^2 + 3\alpha - 2 = 0 \Rightarrow 2\alpha + 3 - \dfrac{2}{\alpha} = 0 \Rightarrow 2\left(\alpha - \dfrac{1}{\alpha}\right) = -3$

        $\therefore \alpha - \dfrac{1}{\alpha} = -\dfrac{3}{2}$

        $\therefore \alpha^2 + \dfrac{1}{\alpha^2} = \left(\alpha - \dfrac{1}{\alpha}\right)^2 + 2 = \dfrac{9}{4} + 2 = 4\dfrac{1}{4}$

(3) $\left(\alpha + \frac{1}{\alpha}\right)^2$ for $x^2 - 3x - 1 = 0$

$\alpha^2 - 3\alpha - 1 = 0 \Rightarrow \alpha - 3 - \frac{1}{\alpha} = 0 \Rightarrow \alpha - \frac{1}{\alpha} = 3$

$\therefore \left(\alpha + \frac{1}{\alpha}\right)^2 = \left(\alpha - \frac{1}{\alpha}\right)^2 + 4 = 13$

(4) $\alpha^2 + \frac{9}{\alpha^2}$ for $3x^2 + 2x - 9 = 0$

$3\alpha^2 + 2\alpha - 9 = 0 \Rightarrow 3\alpha + 2 - \frac{9}{\alpha} = 0 \Rightarrow 3\left(\alpha - \frac{3}{\alpha}\right) = -2 \Rightarrow \alpha - \frac{3}{\alpha} = -\frac{2}{3}$

$\therefore \alpha^2 + \frac{9}{\alpha^2} = \left(\alpha - \frac{3}{\alpha}\right)^2 + 6 = \frac{4}{9} + 6 = 6\frac{4}{9}$

(5) $\frac{\alpha-1}{\alpha+1} - \frac{\alpha+1}{\alpha-1}$ for $2x^2 - 6x - 2 = 0$

$2x^2 - 6x - 2 = 0 \Rightarrow 2(x^2 - 3x - 1) = 0 \Rightarrow x^2 - 3x - 1 = 0 \Rightarrow \alpha^2 - 3\alpha - 1 = 0$

$\therefore \alpha^2 - 1 = 3\alpha$

$\therefore \frac{\alpha-1}{\alpha+1} - \frac{\alpha+1}{\alpha-1} = \frac{(\alpha-1)^2 - (\alpha+1)^2}{(\alpha+1)(\alpha-1)} = \frac{-4\alpha}{\alpha^2-1} = \frac{-4\alpha}{3\alpha} = \frac{-4}{3}$

(6) $\alpha^2 + \alpha - \frac{1}{\alpha} + \frac{1}{\alpha^2}$ for $x^2 - 4x - 1 = 0$

$\alpha^2 - 4\alpha - 1 = 0 \Rightarrow \alpha - 4 - \frac{1}{\alpha} = 0 \Rightarrow \alpha - \frac{1}{\alpha} = 4$

$\therefore \alpha^2 + \frac{1}{\alpha^2} = \left(\alpha - \frac{1}{\alpha}\right)^2 + 2 = 16 + 2 = 18$

$\therefore \alpha^2 + \alpha - \frac{1}{\alpha} + \frac{1}{\alpha^2} = \alpha^2 + \frac{1}{\alpha^2} + \left(\alpha - \frac{1}{\alpha}\right) = 18 + 4 = 22$

#7. **Find the value of the given expression for the following quadratic equations. Each equation has two solutions (α, β) where $\alpha > \beta$.**

(1) $\alpha + \beta$ for $x^2 - 5x - 6 = 0$

$x^2 - 5x - 6 = (x-6)(x+1) = 0$; $x = 6$ or $x = -1$ $\therefore \alpha + \beta = 6 + (-1) = 5$

(2) $\alpha - \beta$ for $x^2 + 3x - 10 = 0$

$x^2 + 3x - 10 = (x+5)(x-2) = 0$; $x = -5$ or $x = 2$ $\therefore \alpha - \beta = 2 - (-5) = 7$

(3) $\frac{\alpha+\beta}{\alpha-\beta}$ for $12x^2 + 5x - 3 = 0$

$12x^2 + 5x - 3 = (3x-1)(4x+3) = 0$; $x = \frac{1}{3}$ or $x = -\frac{3}{4}$

$\therefore \alpha + \beta = \frac{1}{3} + \left(-\frac{3}{4}\right) = \frac{4-9}{12} = -\frac{5}{12}$ and $\alpha - \beta = \frac{1}{3} + \frac{3}{4} = \frac{4+9}{12} = \frac{13}{12}$ $\therefore \frac{\alpha+\beta}{\alpha-\beta} = \frac{-\frac{5}{12}}{\frac{13}{12}} = -\frac{5}{13}$

(4) $\alpha^2 - \beta^2$ for $6x^2 - x - 1 = 0$

$6x^2 - x - 1 = (2x - 1)(3x + 1) = 0$; $x = \frac{1}{2}$ or $x = -\frac{1}{3}$

$\therefore \alpha^2 - \beta^2 = (\alpha + \beta)(\alpha - \beta) = \left(\frac{1}{2} - \frac{1}{3}\right)\left(\frac{1}{2} + \frac{1}{3}\right) = \frac{1}{6} \cdot \frac{5}{6} = \frac{5}{36}$

#8. Solve the following quadratic equations using square roots

(1) $2x^2 = 8$; $x^2 = 4$; $x = \pm 2$

(2) $9x^2 - 5 = 0$; $9x^2 = 5$; $x^2 = \frac{5}{9}$; $x = \pm\sqrt{\frac{5}{9}} = \pm\frac{\sqrt{5}}{3}$

(3) $3(x - 1)^2 = 15$; $(x - 1)^2 = 5$; $x - 1 = \pm\sqrt{5}$; $x = 1 \pm \sqrt{5}$

(4) $(2x + 5)^2 - 3 = 0$; $(2x + 5)^2 = 3$; $2x + 5 = \pm\sqrt{3}$; $x = \frac{-5\pm\sqrt{3}}{2}$

(5) $4(x - 2)^2 - 1 = 0$; $4(x - 2)^2 = 1$; $(x - 2)^2 = \frac{1}{4}$; $x - 2 = \pm\sqrt{\frac{1}{4}}$; $x = 2 \pm \frac{1}{2}$

#9. Solve the following quadratic equations using perfect squares

(1) $x^2 - 3x - 3 = 0$

$\left(x - \frac{3}{2}\right)^2 - \frac{9}{4} - 3 = 0$; $\left(x - \frac{3}{2}\right)^2 = \frac{9}{4} + 3 = \frac{9+12}{4} = \frac{21}{4}$ $\therefore$ $x = \frac{3}{2} \pm \frac{\sqrt{21}}{2}$

(2) $2x^2 + 5x = 7$

$2\left(x^2 + \frac{5}{2}x\right) = 7$; $x^2 + \frac{5}{2}x = \frac{7}{2}$; $\left(x + \frac{5}{4}\right)^2 - \frac{25}{16} = \frac{7}{2}$; $\left(x + \frac{5}{4}\right)^2 = \frac{25+56}{16} = \frac{81}{16}$

$\therefore x = -\frac{5}{4} \pm \frac{9}{4}$ $\therefore x = 1$ or $-\frac{7}{2}$

(3) $-x^2 - 3x + 5 = 0$

$x^2 + 3x - 5 = 0$; $\left(x + \frac{3}{2}\right)^2 - \frac{9}{4} - 5 = 0$; $\left(x + \frac{3}{2}\right)^2 = \frac{9}{4} + 5 = \frac{29}{4}$

$\therefore x = -\frac{3}{2} \pm \frac{\sqrt{29}}{2}$

(4) $3x^2 - 4x + 1 = 0$

$3\left(x^2 - \frac{4}{3}x + \frac{1}{3}\right) = 0$; $x^2 - \frac{4}{3}x + \frac{1}{3} = 0$; $\left(x - \frac{2}{3}\right)^2 - \frac{4}{9} + \frac{1}{3} = 0$; $\left(x - \frac{2}{3}\right)^2 = \frac{4}{9} - \frac{1}{3} = \frac{4-3}{9} = \frac{1}{9}$

$\therefore x = \frac{2}{3} \pm \frac{1}{3}$ $\therefore$ $x = 1$ or $x = \frac{1}{3}$

#10. Find the constant k for the following quadratic equations with a double root:

(1) $(3x - 4)^2 - k^2 = 0$

By formula, $k^2 = 0$; $k = 0$

OR $9x^2 - 24x + 16 - k^2 = 0$; $x^2 - \frac{8}{3}x + \frac{16-k^2}{9} = 0$; $\left(x - \frac{4}{3}\right)^2 - \frac{16}{9} + \frac{16-k^2}{9} = 0$

$\therefore \frac{16}{9} = \frac{16-k^2}{9}$; $k^2 = 0$; $k = 0$

(2) $x^2 - kx + 5 = 0$

By formula, $5 = \left(-\frac{k}{2}\right)^2 = \frac{k^2}{4}$; $k^2 = 20$ $\therefore k = \pm 2\sqrt{5}$

OR $\left(x - \frac{k}{2}\right)^2 - \frac{k^2}{4} + 5 = 0$; $\left(x - \frac{k}{2}\right)^2 = \frac{k^2}{4} - 5$; $\frac{k^2}{4} - 5 = 0$; $\frac{k^2}{4} = 5$; $k^2 = 20$ $\therefore k = \pm 2\sqrt{5}$

(3) $x^2 + 2x + k^2 = 0$

By formula, $k^2 = \left(\frac{2}{2}\right)^2 = 1$ $\therefore k = \pm 1$

OR $(x + 1)^2 - 1 + k^2 = 0$; $(x + 1)^2 = 1 - k^2$; $1 - k^2 = 0$; $k^2 = 1$ $\therefore k = \pm 1$

(4) $kx^2 + 3x + 2 = 0$

By formula, $\frac{2}{k} = \left(\frac{1}{2} \cdot \frac{3}{k}\right)^2 = \frac{9}{4k^2}$ $\therefore 2 = \frac{9}{4k}$ $\therefore k = \frac{9}{8}$

OR $k\left(x^2 + \frac{3}{k}x + \frac{2}{k}\right) = 0$; $\left(x + \frac{3}{2k}\right)^2 - \frac{9}{4k^2} + \frac{2}{k} = 0$, since $k \neq 0$ ($\because$ It's a quadratic equation.)

$\left(x + \frac{3}{2k}\right)^2 = \frac{9}{4k^2} - \frac{2}{k}$ $\therefore \frac{9}{4k^2} = \frac{2}{k}$; $8k^2 = 9k$; $8k = 9$ $\therefore k = \frac{9}{8}$

(5) $2x^2 + 3x + k - 5 = 0$

By formula, $\frac{k-5}{2} = \left(\frac{1}{2} \cdot \frac{3}{2}\right)^2$ $\therefore k - 5 = \frac{9}{16} \cdot 2 = \frac{9}{8}$ $\therefore k = 5 + \frac{9}{8} = 6\frac{1}{8}$

OR $2\left(x^2 + \frac{3}{2}x + \frac{k-5}{2}\right) = 0$; $x^2 + \frac{3}{2}x + \frac{k-5}{2} = 0$; $\left(x + \frac{3}{4}\right)^2 - \frac{9}{16} + \frac{k-5}{2} = 0$

$\left(x + \frac{3}{4}\right)^2 = \frac{9}{16} - \frac{k-5}{2}$; $\frac{k-5}{2} = \frac{9}{16}$; $16(k - 5) = 18$; $8k = 49$; $k = \frac{49}{8} = 6\frac{1}{8}$

(6) $x^2 + kx + (k - 1) = 0$

By formula, $k - 1 = \left(\frac{k}{2}\right)^2 = \frac{k^2}{4}$; $k^2 - 4k + 4 = 0$; $(k - 2)^2 = 0$; $k = 2$

OR $\left(x + \frac{k}{2}\right)^2 - \frac{k^2}{4} + (k - 1) = 0$; $\frac{k^2}{4} = (k - 1)$; $k^2 - 4k + 4 = 0$; $(k - 2)^2 = 0$; $k = 2$

(7) $\frac{1}{3}x^2 + (k+1)x + 8 = 0$

By formula, $24 = \left(\frac{3(k+1)}{2}\right)^2 = \frac{9(k+1)^2}{4}$; $\frac{3(k+1)^2}{4} = 8$; $(k+1)^2 = \frac{32}{3}$

$k = -1 \pm \sqrt{\frac{32}{3}} = -1 \pm \frac{4\sqrt{6}}{3}$

OR $x^2 + 3(k+1)x + 24 = 0$; $\left(x + \frac{3(k+1)}{2}\right)^2 - \frac{9(k+1)^2}{4} + 24 = 0$; $\frac{9(k+1)^2}{4} = 24$

$(k+1)^2 = 24 \cdot \frac{4}{9}$; $k+1 = \pm\sqrt{\frac{32}{3}} = \pm 4\sqrt{\frac{2}{3}} = \pm\frac{4\sqrt{6}}{3}$; $k+1 = \pm\frac{4\sqrt{6}}{3}$ ∴ $k = -1 \pm \frac{4\sqrt{6}}{3}$

#11. Find the value of $p + q$ for the following quadratic equations with the solution $x = p \pm \sqrt{q}$:

(1) $-2x^2 + 5x + 1 = 0$

$-2\left(x^2 - \frac{5}{2}x - \frac{1}{2}\right) = 0$; $x^2 - \frac{5}{2}x - \frac{1}{2} = 0$; $\left(x - \frac{5}{4}\right)^2 - \frac{25}{16} - \frac{1}{2} = 0$; $\left(x - \frac{5}{4}\right)^2 = \frac{25}{16} + \frac{1}{2} = \frac{33}{16}$

∴ $x = \frac{5}{4} \pm \sqrt{\frac{33}{16}}$

So, $p = \frac{5}{4}$, $q = \frac{33}{16}$

Therefore, $p + q = \frac{5}{4} + \frac{33}{16} = \frac{53}{16}$

(2) $3(x-1)^2 = 4$

$(x-1)^2 = \frac{1}{3}$; $x = 1 \pm \sqrt{\frac{4}{3}}$

So, $p = 1$, $q = \frac{4}{3}$

Therefore, $p + q = 1 + \frac{4}{3} = 2\frac{1}{3}$

(3) $-(x+1)^2 + 5 = 0$

$(x+1)^2 = 5$; $x = -1 \pm \sqrt{5}$

So, $p = -1$, $q = 5$

Therefore, $p + q = -1 + 5 = 4$

#12. Find the constant a or the range of a for the following quadratic equations with a condition :

(1) $(x+1)^2 = a + 2$ has no solution.

Since $(x+1)^2 \geq 0$, $a + 2 < 0$ ∴ $a < -2$

(2) $x^2 + 3x + 3a = 0$ **has two different solutions.**

$$\left(x + \frac{3}{2}\right)^2 - \frac{9}{4} + 3a = 0 \; ; \; \left(x + \frac{3}{2}\right)^2 = \frac{9}{4} - 3a \; \therefore \frac{9}{4} - 3a > 0 \; ; \; 3a < \frac{9}{4} \; \therefore \; a < \frac{3}{4}$$

(3) $ax^2 + x + 2 = 0$ **has one solution.**

$$a\left(x^2 + \frac{1}{a}x + \frac{2}{a}\right) = 0$$

Since $a \neq 0$, $x^2 + \frac{1}{a}x + \frac{2}{a} = 0 \; ; \; \left(x + \frac{1}{2a}\right)^2 - \frac{1}{4a^2} + \frac{2}{a} = 0 \; ; \; \frac{1}{4a^2} = \frac{2}{a} \; ; \; \frac{1}{4a} = \frac{2}{1} \; \therefore 8a = 1 \; ; a = \frac{1}{8}$

(4) $3x^2 - x + a = 0$ **has no solution.**

$$3\left(x^2 - \frac{1}{3}x + \frac{a}{3}\right) = 0 \; ; \; x^2 - \frac{1}{3}x + \frac{a}{3} = 0 \; ; \; \left(x - \frac{1}{6}\right)^2 - \frac{1}{36} + \frac{a}{3} = 0 \; ; \; \left(x - \frac{1}{6}\right)^2 = \frac{1}{36} - \frac{a}{3}$$

$$\therefore \; \frac{1}{36} - \frac{a}{3} < 0 \; ; \; \frac{a}{3} > \frac{1}{36} \; \therefore \; a > \frac{1}{12}$$

(5) $x^2 + (a + 1)x + \frac{a+3}{2}$ **has a double root.**

$$\left(x + \frac{a+1}{2}\right)^2 - \frac{(a+1)^2}{4} + \frac{a+3}{2} = 0 \; ; \; \frac{(a+1)^2}{4} = \frac{a+3}{2} \; ; \; (a+1)^2 = 2a + 6 \; ; \; a^2 + 2a + 1 = 2a + 6$$

$$a^2 = 5 \quad \therefore \; a = \pm\sqrt{5}$$

#13. Find the value of the given expression for the following quadratic equations with two solutions (α, β) :

(1) $\alpha\beta$ **for** $2(x + 3)^2 - 3 = 0$

$$2(x + 3)^2 = 3 \; ; \; (x + 3)^2 = \frac{3}{2} \; ; \; x = -3 \pm \sqrt{\frac{3}{2}}$$

$$\therefore \; \alpha\beta = \left(-3 + \sqrt{\frac{3}{2}}\right)\left(-3 - \sqrt{\frac{3}{2}}\right) = 9 - \frac{3}{2} = 7\frac{1}{2}$$

(2) $\alpha + \beta$ **for** $-(x + 4)^2 + 5 = 0$

$$(x + 4)^2 = 5 \; ; \; x = -4 \pm \sqrt{5}$$

$$\therefore \; \alpha + \beta = \left(-4 + \sqrt{5}\right) + \left(-4 - \sqrt{5}\right) = -8$$

(3) $\alpha^2 + \beta^2$ **for** $3x^2 - x - 1 = 0$

$$3\left(x^2 - \frac{1}{3}x - \frac{1}{3}\right) = 0 \; ; \; x^2 - \frac{1}{3}x - \frac{1}{3} = 0 \; ; \; \left(x - \frac{1}{6}\right)^2 - \frac{1}{36} - \frac{1}{3} = 0 \; ; \; \left(x - \frac{1}{6}\right)^2 = \frac{1}{36} + \frac{1}{3} = \frac{1+12}{36} = \frac{13}{36}$$

$$\therefore x = \frac{1}{6} \pm \frac{\sqrt{13}}{6}$$

So, $\alpha + \beta = \frac{2}{6} = \frac{1}{3}$ and $\alpha\beta = \left(\frac{1}{6} + \frac{\sqrt{13}}{6}\right)\left(\frac{1}{6} - \frac{\sqrt{13}}{6}\right) = \frac{1}{36} - \frac{13}{36} = \frac{-12}{36} = -\frac{1}{3}$

$$\therefore \; \alpha^2 + \beta^2 = (\alpha + \beta)^2 - 2\alpha\beta = \frac{1}{9} + \frac{2}{3} = \frac{7}{9}$$

#14. Find two constants (a and b) for the following quadratic equations with a condition :

(1) $x^2 + x + a = 0$ **has two different solutions,** $x = 2$ **and** $x = b$.

$x^2 + x + a = (x - 2)(x - b) = x^2 + (-2 - b)x + 2b = 0$

$\therefore -2 - b = 1$ and $2b = a$

$\therefore b = -3$ and $a = -6$

(2) $x^2 + 2ax + b = 0$ **has a double root** $x = 3$.

$x^2 + 2ax + b = (x + a)^2 - a^2 + b = (x - 3)^2 = 0$

$\therefore a = -3$ and $-a^2 + b = 0$; $b = 9$

(3) $x^2 + ax + b = 0$ **has a solution** $x = 1 + \sqrt{2}$.

Since the other solution is $x = 1 - \sqrt{2}$,

$x^2 + ax + b = \left(x - \left(1 + \sqrt{2}\right)\right)\left(x - \left(1 - \sqrt{2}\right)\right)$

$= x^2 - \left(1 + \sqrt{2}\right)x - \left(1 - \sqrt{2}\right)x + (1 - 2) = x^2 - 2x - 1 = 0$

$\therefore a = -2$ and $b = -1$

OR $\left(1 + \sqrt{2}\right)^2 + a\left(1 + \sqrt{2}\right) + b = 0$; $1 + 2\sqrt{2} + 2 + a + a\sqrt{2} + b = 0$

$1 + 2 + a + b = 0$ and $2\sqrt{2} + a\sqrt{2} = 0$

$\therefore a = -2$ and $b = -1$

OR $\left(x + \frac{a}{2}\right)^2 - \frac{a^2}{4} + b = 0$; $x = -\frac{a}{2} \pm \sqrt{\frac{a^2 - 4b}{4}}$

$\therefore -\frac{a}{2} = 1, \frac{a^2 - 4b}{4} = 2$; $a = -2$ and $b = -1$

(4) $x^2 - ax - 2b^2 = 0$ **has two different solutions,** $x = 4 \pm \sqrt{2a}$.

$\left(x - \frac{a}{2}\right)^2 - \frac{a^2}{4} - 2b^2 = 0$; $\left(x - \frac{a}{2}\right)^2 = \frac{a^2}{4} + 2b^2$; $x = \frac{a}{2} \pm \sqrt{\frac{a^2 + 8b^2}{4}}$

$\therefore \frac{a}{2} = 4, \frac{a^2 + 8b^2}{4} = 2a$; $a = 8$, $a^2 + 8b^2 = 8a$; $8b^2 = 64 - 64 = 0$; $b = 0$

$\therefore a = 8$ and $b = 0$

#15. Solve the following quadratic equations using quadratic formulas

(1) $x^2 - 2x - 4 = 0$

$x = \frac{1 \pm \sqrt{1 + 4}}{1} = 1 \pm \sqrt{5}$

(2) $3x^2 + 5x - 1 = 0$

$$x = \frac{-5 \pm \sqrt{25 - 4 \cdot 3 \cdot (-1)}}{2 \cdot 3} = \frac{-5 \pm \sqrt{37}}{6}$$

(3) $5x^2 - 2x - 1 = 0$

$$x = \frac{1 \pm \sqrt{1 + 5}}{5} = \frac{1 \pm \sqrt{6}}{5}$$

(4) $-2x^2 + 3x + 5 = 0$

$$x = \frac{-3 \pm \sqrt{9 - 4 \cdot (-2) \cdot 5}}{2 \cdot (-2)} = \frac{-3 \pm \sqrt{9 + 40}}{-4} = \frac{-3 \pm 7}{-4} \quad ; \quad x = -1 \text{ or } x = \frac{5}{2}$$

(5) $\frac{1}{2}x^2 - 3x + 2 = 0$

$$x^2 - 6x + 4 = 0 \quad ; \quad x = \frac{3 \pm \sqrt{9 - 4}}{1} = 3 \pm \sqrt{5}$$

(6) $\frac{1}{6}x^2 - 0.5x + \frac{1}{4} = 0$

$$2x^2 - 6x + 3 = 0 \quad ; \quad x = \frac{3 \pm \sqrt{9 - 6}}{2} = \frac{3 \pm \sqrt{3}}{2}$$

(7) $(x + 1)^2 = 3(x + 2)$

$$x^2 + 2x + 1 - 3x - 6 = 0 \quad ; \quad x^2 - x - 5 = 0 \quad ; \quad x = \frac{1 \pm \sqrt{1 + 20}}{2 \cdot 1} = \frac{1 \pm \sqrt{21}}{2}$$

(8) $(x + 2)^2 + 3(x + 2) - 2 = 0$

Letting $A = x + 2$, $A^2 + 3A - 2 = 0$

$$A = \frac{-3 \pm \sqrt{9 + 8}}{2} = \frac{-3 \pm \sqrt{17}}{2} \quad ; \quad x + 2 = \frac{-3 \pm \sqrt{17}}{2} \quad ; \quad x = \frac{-4 - 3 \pm \sqrt{17}}{2} = \frac{-7 \pm \sqrt{17}}{2}$$

(9) $-\frac{(x-1)^2}{2} + x = 0.4(x + 1)$

$$-5(x - 1)^2 + 10x = 4(x + 1); \quad 5(x - 1)^2 + 4(x + 1) - 10x = 0 \quad ; \quad 5x^2 - 16x + 9 = 0$$

$$x = \frac{8 \pm \sqrt{64 - 45}}{5} = \frac{8 \pm \sqrt{19}}{5}$$

(10) $(x + 3)(2x + 6) = 5$

Substituting $A = x + 3$, $A \cdot 2A - 5 = 0$; $2A^2 - 5 = 0$

$$A = \frac{-0 \pm \sqrt{0 + 40}}{2 \cdot 2} = \frac{\pm \sqrt{40}}{4} = \frac{\pm 2\sqrt{10}}{4} = \frac{\pm \sqrt{10}}{2} \quad ; \quad x + 3 = \frac{\pm \sqrt{10}}{2} \quad ; \quad x = -3 \pm \frac{\sqrt{10}}{2}$$

OR $(x + 3)(2x + 6) - 5 = 2x^2 + 12x + 13 = 0$

$$x = \frac{-6 \pm \sqrt{36 - 26}}{2} = \frac{-6 \pm \sqrt{10}}{2} = -3 \pm \frac{\sqrt{10}}{2}$$

#16. Find the value of the given expression for the following quadratic equations with a solution

(1) $a + b$ for $(2x+1)^2 = 3$ with $x = a \pm b\sqrt{3}$

$4x^2 + 4x - 2 = 0$; $2x^2 + 2x - 1 = 0$; $x = \frac{-1\pm\sqrt{1+2}}{2} = \frac{-1\pm\sqrt{3}}{2}$

$\therefore a = \frac{-1}{2}, b = \frac{1}{2}$ $\therefore a + b = 0$

(2) $a - b$ for $2x^2 - 8x + 1 = 0$ with $x = \frac{a\pm3\sqrt{b}}{6}$

$2x^2 - 8x + 1 = 0$; $x = \frac{4\pm\sqrt{16-2}}{2} = \frac{4\pm\sqrt{14}}{2} = \frac{12\pm3\sqrt{14}}{6}$

$\therefore a = 12, b = 14$ $\therefore a - b = -2$

(3) ab for $ax^2 + 5x + 2 = 0$ with $x = \frac{-5\pm2\sqrt{b}}{4}$

$x = \frac{-5\pm\sqrt{25-8a}}{2a}$ $\therefore$ $\frac{-5}{2a} = \frac{-5}{4}$; $a = 2$

$\therefore$ $\frac{\sqrt{25-8a}}{2a} = \frac{2\sqrt{b}}{4}$; $\frac{\sqrt{25-16}}{4} = \frac{2\sqrt{b}}{4}$; $\sqrt{9} = 2\sqrt{b}$; $9 = 4b$; $b = \frac{9}{4}$

$\therefore ab = \frac{9}{2}$

(4) $\frac{b}{a}$ for $3x^2 - 5x + 1 = 0$ with $x = a \pm \sqrt{b}$

$x = \frac{5\pm\sqrt{25-12}}{6} = \frac{5\pm\sqrt{13}}{6}$

$\therefore$ $a = \frac{5}{6}$, $b = \frac{13}{36}$

$\therefore$ $\frac{b}{a} = \frac{13}{36} \cdot \frac{6}{5} = \frac{13}{30}$

(5) $\frac{a+b}{ab}$ for $x^2 - 3x + 1 = 0$ with $x = \frac{a\pm2\sqrt{b}}{2}$

$x = \frac{3\pm\sqrt{9-4}}{2} = \frac{3\pm\sqrt{5}}{2}$

$\therefore$ $a = 3$, $4b = 5$; $b = \frac{5}{4}$

Since $a + b = 3 + \frac{5}{4} = \frac{17}{4}$ and $ab = 3 \cdot \frac{5}{4} = \frac{15}{4}$, $\frac{a+b}{ab} = \frac{17}{4} \cdot \frac{4}{15} = \frac{17}{15}$

(6) $\frac{a-b}{a^2-b^2}$ for $ax^2 + 3x - 3b = 0$ with $x = -1 \pm \sqrt{5}$

$$x = \frac{-3 \pm \sqrt{9+12ab}}{2a} = -1 \pm \sqrt{5} \quad \therefore \frac{-3}{2a} = -1 ; \quad a = \frac{3}{2}$$

$$\frac{\sqrt{9+12ab}}{2a} = \frac{\sqrt{9+18b}}{3} = \sqrt{\frac{9+18b}{9}} = \sqrt{1+2b} = \sqrt{5} \quad ; 1 + 2b = 5 \; ; \; b = 2$$

$$\therefore \frac{a-b}{a^2-b^2} = \frac{a-b}{(a+b)(a-b)} = \frac{1}{(a+b)} = \frac{1}{\frac{3}{2}+2} = \frac{2}{7}$$

#17. Identify the number of solutions for each quadratic equation.

(1) $x^2 + 2x - 3 = 0$

$D = 4 - 4 \cdot 1 \cdot (-3) = 4 + 12 = 16 > 0 \quad \therefore 2$ different solutions.

(2) $-x^2 + x - 5 = 0$

$D = 1 - 4 \cdot (-1) \cdot (-5) = 1 - 20 = -19 < 0 \quad \therefore$ no solution.

(3) $4x^2 - 4x + 1 = 0$

$D = 16 - 4 \cdot 4 \cdot 1 = 0 \quad \therefore$ only one solution.

(4) $kx^2 - (k+5)x + 1 = 0$

$D = (k+5)^2 - 4 \cdot k \cdot 1 = k^2 + 6k + 25 = (k+3)^2 - 9 + 25$

$= (k+3)^2 + 16 > 0 \; (\because (k+3)^2 \geq 0) \quad \therefore 2$ different solutions.

(5) $3x^2 - x - k^2 = 0$

$D = 1 + 12k^2$

Since $k^2 \geq 0$, $D = 1 + 12k^2 > 0 \quad \therefore 2$ different solutions.

(6) $x^2 - 4kx + 5k^2 + 1 = 0$

$D = 16k^2 - 4(5k^2 + 1) = -4k^2 - 4 = -4(k^2 + 1) < 0 \; (\because k^2 + 1 > 0)$

$\therefore$ no solution.

#18. Find the value of a or range of a for the following quadratic equations with a condition (Use the discriminant D)

(1) $x^2 + 5x + a = x + 2$ has no solution.

$x^2 + 4x + a - 2 = 0$

$D = 16 - 4(a-2) < 0 \quad \therefore 4(a-2) > 16 ; \; a - 2 > 4 ; \; a > 6$

(2) $(a+3)x^2 - 2ax + a - 1 = 0$ **has two different solutions.**

$D = 4a^2 - 4(a+3)(a-1) > 0$; $-8a + 12 > 0$; $a < \frac{3}{2}$

Since $a + 3 \neq 0$, $a \neq -3$

∴ $a < -3$ or $-3 < a < \frac{3}{2}$

(3) $x^2 + 3ax - 2a + 3 = 0$ **has only one solution.**

$D = (3a)^2 - 4 \cdot 1 \cdot (-2a + 3) = 0$; $9a^2 + 8a - 12 = 0$

∴ $a = \frac{-4 \pm \sqrt{16 + 9 \cdot 12}}{9} = \frac{-4 \pm 2\sqrt{31}}{9}$

(4) $x^2 + ax + a + 2 = 0$ **has a double root and** $x^2 + 4ax + (2a-1)^2 = 0$ **has two different solutions.**

Since $x^2 + ax + a + 2 = 0$ has a double root, $D = a^2 - 4(a+2) = 0$

$a^2 - 4a - 8 = 0$; $(a-2)^2 - 4 - 8 = 0$; $(a-2)^2 = 12$ ∴ $a = 2 + 2\sqrt{3}$ or $a = 2 - 2\sqrt{3}$

Since $x^2 + 4ax + (2a-1)^2 = 0$ has two different solutions, $D = (4a)^2 - 4 \cdot 1 \cdot (2a-1)^2 > 0$

$16a^2 - 4(4a^2 - 4a + 1) > 0$; $16a - 4 > 0$; $a > \frac{1}{4}$

Therefore, $a = 2 + 2\sqrt{3}$

(5) $2x^2 + (2a-1)x + a^2 + \frac{1}{4} = 0$ **has solutions.**

$D \geq 0$

$D = (2a-1)^2 - 4 \cdot 2 \cdot (a^2 + \frac{1}{4}) \geq 0$; $4a^2 - 4a + 1 - 8a^2 - 2 \geq 0$; $-4a^2 - 4a - 1 \geq 0$

$4a^2 + 4a + 1 \leq 0$; $(2a+1)^2 \leq 0$

Since $(2a+1)^2 \geq 0$, $(2a+1)^2 = 0$

∴ $2a + 1 = 0$; $a = -\frac{1}{2}$

(6) $(a-1)x^2 + 2(a-1)x + (a+1) = 0$ **has solutions.**

Since this equation is quadratic, $a - 1 \neq 0$; $a \neq 1$

To have solutions, $D \geq 0$

$D = 4(a-1)^2 - 4 \cdot (a-1) \cdot (a+1) \geq 0$; $a^2 - 2a + 1 - (a^2 - 1) \geq 0$

$-2a + 2 \geq 0$; $2a \leq 2$; $a \leq 1$

Since $a \neq 1$, $a < 1$

#19. A quadratic equation $3x^2 + 5x - 2 = 0$ has two solutions, $x = \alpha$ and $x = \beta$. Find the value of the given expressions.

(1) $\alpha + \beta$

$\alpha + \beta = -\dfrac{5}{3}$

(2) $\alpha^2 + \beta^2$

$\alpha^2 + \beta^2 = (\alpha + \beta)^2 - 2\alpha\beta = \left(-\dfrac{5}{3}\right)^2 - 2\left(-\dfrac{2}{3}\right) = \dfrac{37}{9}$

(3) $\alpha - \beta$

$(\alpha - \beta)^2 = (\alpha + \beta)^2 - 4\alpha\beta = \left(-\dfrac{5}{3}\right)^2 - 4\left(-\dfrac{2}{3}\right) = \dfrac{49}{9}$; $\alpha - \beta = \pm\dfrac{7}{3}$

(4) $\alpha^2 - \beta^2$

$\alpha^2 - \beta^2 = (\alpha + \beta)(\alpha - \beta) = \begin{cases} -\dfrac{35}{9}, & \text{when } \alpha - \beta = \dfrac{7}{3} \\ \dfrac{35}{9}, & \text{when } \alpha - \beta = -\dfrac{7}{3} \end{cases}$

(5) $\dfrac{1}{\alpha} + \dfrac{1}{\beta}$

$\dfrac{1}{\alpha} + \dfrac{1}{\beta} = \dfrac{\alpha + \beta}{\alpha\beta} = \dfrac{-\frac{5}{3}}{-\frac{2}{3}} = \dfrac{5}{2}$

#20. A quadratic equation $x^2 + 3kx + 2k^2 - 4k - 1 = 0$ has two solutions, $x = \alpha$ and $x = \beta$. Find the value of the given expressions in terms of k for (1) through (5). Find the value of k for (6).

(1) $\alpha + \beta$; $\alpha + \beta = -\dfrac{3k}{1} = -3k$

(2) $\alpha\beta$; $\alpha\beta = \dfrac{2k^2 - 4k - 1}{1} = 2k^2 - 4k - 1$

(3) $\alpha^2 + \beta^2$; $\alpha^2 + \beta^2 = (\alpha + \beta)^2 - 2\alpha\beta = 9k^2 - 4k^2 + 8k + 2 = 5k^2 + 8k + 2$

(4) $(\alpha - \beta)^2$; $(\alpha - \beta)^2 = (\alpha + \beta)^2 - 4\alpha\beta = 9k^2 - 8k^2 + 16k + 4 = k^2 + 16k + 4$

(5) $\dfrac{\beta}{\alpha} + \dfrac{\alpha}{\beta}$; $\dfrac{\beta}{\alpha} + \dfrac{\alpha}{\beta} = \dfrac{\alpha^2 + \beta^2}{\alpha\beta} = \dfrac{5k^2 + 8k + 2}{2k^2 - 4k - 1}$

(6) k if $\dfrac{1}{\alpha} + \dfrac{1}{\beta} = 1$; $\dfrac{1}{\alpha} + \dfrac{1}{\beta} = \dfrac{\alpha + \beta}{\alpha\beta} = \dfrac{-3k}{2k^2 - 4k - 1} = 1$; $2k^2 - 4k - 1 = -3k$; $2k^2 - k - 1 = 0$

$(2k + 1)(k - 1) = 0$; $k = -\dfrac{1}{2}$ or $k = 1$

#21. Find the solution for the quadratic equation $ax^2 + (b-1)x + 4 = 0$

(1) When the quadratic equation $2x^2 + (a-1)x + b = 0$ has two solutions, $\frac{1}{2}$ and $\frac{1}{3}$.

$\frac{1}{2} + \frac{1}{3} = -\frac{a-1}{2}$; $\frac{5}{6} = -\frac{3(a-1)}{6}$; $a - 1 = -\frac{5}{3}$; $a = -\frac{2}{3}$

$\frac{1}{2} \cdot \frac{1}{3} = \frac{b}{2}$; $b = \frac{1}{3}$

(OR $2\left(x - \frac{1}{2}\right)\left(x - \frac{1}{3}\right) = 0$; $2\left(x^2 - \frac{5}{6}x + \frac{1}{6}\right) = 0$

$2x^2 - \frac{5}{3}x + \frac{1}{3} = 0$; $a - 1 = -\frac{5}{3}$; $a = -\frac{2}{3}$, $b = \frac{1}{3}$)

$\therefore \ ax^2 + (b-1)x + 4 = -\frac{2}{3}x^2 - \frac{2}{3}x + 4 = 0$; $x^2 + x - 6 = 0$; $(x+3)(x-2) = 0$

$\therefore \ x = -3$, or $x = 2$

(2) When the quadratic equation $3ax^2 + 8bx + 3 = 0$ has a double root -2.

$-2 + (-2) = -\frac{8b}{3a}$, $-2 \cdot (-2) = \frac{3}{3a}$

$\therefore \ -4 = -\frac{8b}{3a}$, $4 = \frac{1}{a}$; $a = \frac{1}{4}$, $b = \frac{3}{8}$

(OR $3a(x+2)^2 = 0$; $3a(x^2 + 4x + 4) = 0$; $3ax^2 + 12ax + 12a = 0$; $8b = 12a$, $3 = 12a$

$a = \frac{1}{4}$, $b = \frac{3}{8}$)

$\cdot \ ax^2 + (b-1)x + 1 = \frac{1}{4}x^2 - \frac{5}{8}x + 1 = 0$; $2x^2 - 5x + 32 = 0$

Since $D = 25 - 4 \cdot 2 \cdot 32 < 0$, no solution.

(3) When the quadratic equation $ax^2 + 3ax - 4 = 0$ has two solutions, b and $b + 1$.

$b + (b+1) = -\frac{3a}{a} = -3$; $2b = -4$; $b = -2$

$b(b+1) = \frac{-4}{a}$; $-2(-2+1) = \frac{-4}{a}$; $2a = -4$; $a = -2$

(OR $a(x-b)\big(x-(b+1)\big) = 0$; $a\big(x^2 - (b+b+1)x + b(b+1)\big) = 0$

; $ax^2 - (2b+1)ax + ab(b+1) = 0$ $\therefore \ -(2b+1) = 3$, $ab(b+1) = -4$

$-2b = 4$; $b = -2$ and $ab(b+1) = -4$; $a = -2$)

$\therefore \ ax^2 + (b-1)x + 4 = -2x^2 - 3x + 4$

$\therefore \ x = \frac{3 \pm \sqrt{9 - 4 \cdot (-2) \cdot 4}}{2 \cdot (-2)} = \frac{3 \pm \sqrt{41}}{-4}$

(4) When the quadratic equation $ax^2 + 3x + b = 0$ has two solutions, α and β, which satisfy the conditions $\alpha + \beta = -2$ and $\alpha\beta = 4$.

$\alpha + \beta = -\dfrac{3}{a} = -2$; $a = \dfrac{3}{2}$

$\alpha\beta = \dfrac{b}{a} = 4$; $b = 4a = 4 \cdot \dfrac{3}{2} = 6$

(OR $ax^2 + 3x + b = a(x^2 - (\alpha + \beta)x + \alpha\beta) = a(x^2 - (-2)x + 4) = 0$

$3 = -a(-2)$, $b = a \cdot (4)$; $a = \dfrac{3}{2}, b = 6$)

$\therefore ax^2 + (b-1)x + 4 = \dfrac{3}{2}x^2 + 5x + 4 = 0$

$3x^2 + 10x + 8 = 0$ $\quad \therefore x = \dfrac{-5 \pm \sqrt{25-24}}{3} = \dfrac{-5 \pm 1}{3}$ $\quad \therefore x = -\dfrac{4}{3}$ or $x = -2$

(5) When the quadratic equation $x^2 + ax + 3 = 0$ has two different solutions. The one of the solutions is $x = -2 + 3\sqrt{b}$.

The other solution is $x = -2 - 3\sqrt{b}$.

$\therefore -\dfrac{a}{1} = \left(-2 + 3\sqrt{b}\right) + \left(-2 - 3\sqrt{b}\right) = -4$; $a = 4$

and $\dfrac{3}{1} = \left(-2 + 3\sqrt{b}\right) \cdot \left(-2 - 3\sqrt{b}\right) = 4 - 9b$; $b = \dfrac{1}{9}$

(OR $\left(x - (-2 + 3\sqrt{b})\right)\left(x - (-2 - 3\sqrt{b})\right) = 0$

$x^2 - \left(-2 + 3\sqrt{b} - 2 - 3\sqrt{b}\right)x + \left(-2 + 3\sqrt{b}\right)\left(-2 - 3\sqrt{b}\right) = 0$

$x^2 + 4x + 4 - 9b = 0$ $\therefore a = 4$, $3 = 4 - 9b$ $\therefore a = 4$, $b = \dfrac{1}{9}$)

Therefore, $ax^2 + (b-1)x + 4 = 4x^2 - \dfrac{8}{9}x + 4 = 0$; $36x^2 - 8x + 36 = 0$

$9x^2 - 2x + 9 = 0$.

Since $D = 4 - 4 \cdot 9 \cdot 9 < 0$, no solution.

#22. $x^2 + ax + b = 0$ has two solutions, -2 and -3. Find the value of $\alpha^2 + \beta^2$ for $x^2 - bx - a = 0$ which has solutions, α, β.

$-2 + (-3) = -\dfrac{a}{1}$; $a = 5$ and $-2 \cdot (-3) = \dfrac{b}{1}$; $b = 6$

(OR $(x+2)(x+3) = 0$; $x^2 + 5x + 6 = 0$ $\therefore a = 5, b = 6$)

$\therefore x^2 - bx - a = x^2 - 6x - 5 = 0$; $\alpha + \beta = 6$, $\alpha\beta = -5$

$\therefore \alpha^2 + \beta^2 = (\alpha + \beta)^2 - 2\alpha\beta = 6^2 + 10 = 46$

#23. Create a 1 x^2-coefficient quadratic equation which has two solutions $\alpha + \beta$ and $\alpha\beta$, where α and β are both solutions of $x^2 + 2x - 3 = 0$.

$\alpha + \beta = -2, \ \alpha\beta = -3 \ \therefore \ x^2 - (\alpha + \beta + \alpha\beta)x + (\alpha + \beta) \cdot (\alpha\beta) = 0 \ \ \therefore \ x^2 + 5x + 6 = 0$

(OR $(x - (\alpha + \beta))(x - \alpha\beta) = 0 \ ; \ (x + 2)(x + 3) = 0 \ \therefore \ x^2 + 5x + 6 = 0$)

#24. Find the range of a for the following quadratic equations with a condition

(1) The quadratic equation $x^2 - 3x + 2a = 0$ has two different positive solutions.

Since the equation has two different solutions, $D > 0$

$\therefore D = 9 - 4 \cdot 1 \cdot 2a > 0 \ ; \ 8a < 9 \ ; \ a < \dfrac{9}{8}$

Let α, β be the solutions. Then, $\alpha + \beta = 3$, $\alpha\beta = 2a$

Since α and β are positive, $\alpha\beta = 2a > 0 \ ; \ a > 0$

Therefore, $0 < a < \dfrac{9}{8}$

(2) The quadratic equation $ax^2 + 2x + 3 = 0$ has two different negative solutions.

Since the equation has two different solutions, $D > 0$

$\therefore D = 4 - 4 \cdot a \cdot 3 > 0 \ ; 4 > 12a \ ; \ a < \dfrac{1}{3}$

Let α, β be the solutions. Then, $\alpha + \beta = -\dfrac{2}{a}$, $\alpha\beta = \dfrac{3}{a}$

Since α and β are negative, $\alpha + \beta = -\dfrac{2}{a} < 0 \ ; \ a > 0$ and $\alpha\beta = \dfrac{3}{a} > 0 \ ; \ a > 0$

Therefore, $0 < a < \dfrac{1}{3}$

(3) The quadratic equation $x^2 - 4x + 3a = 0$ has two different solutions α and β with opposite signs.

$\alpha\beta = 3a < 0 \ ; \ a < 0$

$\therefore D = 16 - 4 \cdot 1 \cdot 3a > 0 \ ; \ 16 > 12a \ ; \ a < \dfrac{4}{3}$

Therefore, $a < 0$

#25. The following quadratic equations have only one solution. Find the solution (double root) for each.

(1) $x^2 + kx + 2k - 3 = 0$

$D = k^2 - 4(2k - 3) = 0 \ ; \ k^2 - 8k + 12 = 0 \ ; \ (k - 6)(k - 2) = 0 \ \therefore \ k = 6$ or $k = 2$

If $k = 6$, then $x^2 + kx + 2k - 3 = x^2 + 6x + 9 = 0 \ \therefore \ (x + 3)^2 = 0 \ ; \ x = -3$ (double root)

If $k = 2$, then $x^2 + kx + 2k - 3 = x^2 + 2x + 1 = 0 \ \therefore \ (x + 1)^2 = 0 \ ; \ x = -1$ (double root)

(OR by formula, $x = \frac{-b \pm \sqrt{b^2 - 4ac}}{2a}$

Since $D = 0$, $b^2 - 4ac = 0$ $\therefore x = \frac{-b}{2a}$

So, $x = \frac{-b}{2a} = \frac{-6}{2 \cdot 1} = -3$ (when $k = 6$) or $x = \frac{-b}{2a} = \frac{-2}{2 \cdot 1} = -1$ (when $k = 2$))

(2) $(k + 2)x^2 - 2kx + k + 1 = 0$

$D = 4k^2 - 4(k + 2)(k + 1) = 0$; $4k^2 - 4k^2 - 12k - 8 = 0$ $\therefore k = -\frac{2}{3}$

$\therefore (k + 2)x^2 - 2kx + k + 1 = \frac{4}{3}x^2 + \frac{4}{3}x + \frac{1}{3} = 0$; $4x^2 + 4x + 1 = 0$; $(2x + 1)^2 = 0$

$\therefore x = -\frac{1}{2}$ (double root)

(OR by formula, $x = \frac{-b}{2a} = \frac{-\frac{4}{3}}{2 \cdot \frac{4}{3}} = -\frac{1}{2}$)

(3) $x^2 + (k + 2)x + k^2 - k + 2 = 0$

$D = (k + 2)^2 - 4(k^2 - k + 2) = 0$; $k^2 + 4k + 4 - 4k^2 + 4k - 8 = 0$

$-3k^2 + 8k - 4 = 0$; $3k^2 - 8k + 4 = 0$; $(k - 2)(3k - 2) = 0$

$\therefore k = 2$ or $k = \frac{2}{3}$

If $k = 2$, then $x^2 + (k + 2)x + k^2 - k + 2 = x^2 + 4x + 4 = 0$

$\therefore (x + 2)^2 = 0$ $\therefore x = -2$ (double root)

If $k = \frac{2}{3}$, then $x^2 + \frac{8}{3}x + \frac{4}{9} - \frac{2}{3} + 2 = 0$; $x^2 + \frac{8}{3}x + \frac{16}{9} = 0$; $9x^2 + 24x + 16 = 0$

$\therefore x = \frac{-12 \pm \sqrt{144 - 9 \cdot 16}}{9} = \frac{-12}{9} = -\frac{4}{3}$ (double root)

#26. An n-sided polygon has $\frac{n(n-3)}{2}$ diagonals. Find a polygon that has 20 diagonals.

$\frac{n(n-3)}{2} = 20$; $n^2 - 3n - 40 = 0$; $(n - 8)(n + 5) = 0$; $n = 8$ or $n = -5$

Since $n > 0$, $n = 8$

Therefore, 8-sided polygon.

#27. For three consecutive positive integers, the square of the biggest number is 12 less than the sum of the squares of the other numbers. Identify the biggest number.

Let n, $n + 1$, $n + 2$ be the consecutive positive numbers. Then

$(n + 2)^2 = n^2 + (n + 1)^2 - 12$; $n^2 + 4n + 4 = 2n^2 + 2n - 11$; $n^2 - 2n - 15 = 0$

$(n - 5)(n + 3) = 0$ $\therefore n = 5$ or $n = -3$

Since $n > 0$, $n = 5$ Therefore, the biggest number is 7.

#28. **The product of two consecutive odd numbers is 99. Find the sum of the numbers.**

Let $2n - 1$, $2n + 1$ be the two consecutive odd numbers. Then,

$(2n - 1)(2n + 1) = 99$, $n \geq 1$

So, $4n^2 - 1 = 99$; $4n^2 = 100$; $n^2 = 25$

Since $n \geq 1$, $n = 5$

Therefore, the sum is $9 + 11 = 20$.

#29. **The sum of two positive numbers is 34 and their product is 225. Identify the two numbers.**

Let x and y be the two positive numbers. Then,

$x + y = 34$ and $xy = 225$

So, $x(34 - x) = 225$; $x^2 - 34x + 225 = 0$

$\therefore \quad x = \frac{17 \pm \sqrt{17^2 - 225}}{1} = 17 \pm \sqrt{64} = 17 \pm 8$

Therefore, the two numbers are 25 and 9

#30. **Nichole wants to produce a x^2 % of salt solution after mixing 40 ounces of a 10% of salt solution with 40 ounces of a x% salt solution. Find the value of x.**

$40 \cdot \frac{10}{100} + 40 \cdot \frac{x}{100} = 80 \cdot \frac{x^2}{100}$

$400 + 40x = 80\, x^2$; $2x^2 - x - 10 = 0$; $(2x - 5)(x + 2) = 0$; $x = \frac{5}{2}$ or $x = -2$

Since $x > 0$, $x = \frac{5}{2}$

#31. **The difference between two positive integers is 2 and their product is 255. Find the sum of the numbers.**

$x - y = 2$, $xy = 255$; $x(x - 2) = 255$; $x^2 - 2x - 255 = 0$; $(x - 17)(x + 15) = 0$

So, $x = 17$ and $y = 15$

Therefore, $x + y = 32$

(OR $x = \frac{1 \pm \sqrt{1 + 255}}{1} = 1 \pm \sqrt{256} = 1 \pm 16$; $x = 17$ or $x = -15$. Since $x > 0$, $x = 17$

$\therefore$ The sum of the numbers is $17 + 15 = 32$.)

#32. The area of a square A is 121 square inches. Each side of square A is 3 inches longer than that of square B. Find the perimeter of the Square B.

Let x be the length of one side of square B. Then,

$(x + 3)^2 = 121$; $x^2 + 6x + 9 - 121 = 0$; $x^2 + 6x - 112 = 0$

So, $x = \frac{-3 \pm \sqrt{9+112}}{1} = -3 \pm \sqrt{121} = -3 \pm 11$; $x = 8$ or $x = -14$

Since $x > 0$, $x = 8$

Therefore, the perimeter of the square B is $4 \cdot 8 = 32$ inches.

#33. The perimeter and the area of a rectangle are 26 inches and 40 square inches, respectively. Find the difference between the length and width of the rectangle's sides (in this case, length will be longer than width).

Let x be the length and y be the width. Then,

$2(x + y) = 26$; $x + y = 13$ and $xy = 40$

Since $(x - y)^2 = (x + y)^2 - 4xy = 13^2 - 4 \cdot 40 = 169 - 160 = 9$, $x - y = \pm 3$

Since $x > y$, $x - y = 3$

Therefore, the difference between the length and width is 3 inches.

(OR $x(13 - x) = 40$; $x^2 - 13x + 40 = 0$; $(x - 5)(x - 8) = 0$; $x = 5$ or $x = 8$

If $x = 5$, then $y = 8$. If $x = 8$, then $y = 5$ So, the difference is $8 - 5 = 3$.)

#34. Richard throws a ball upward with a beginning speed v of 60 feet per second. The formula for the height in feet after t seconds is $h = vt - 5t^2$

(1) At what time will the height 100 feet?

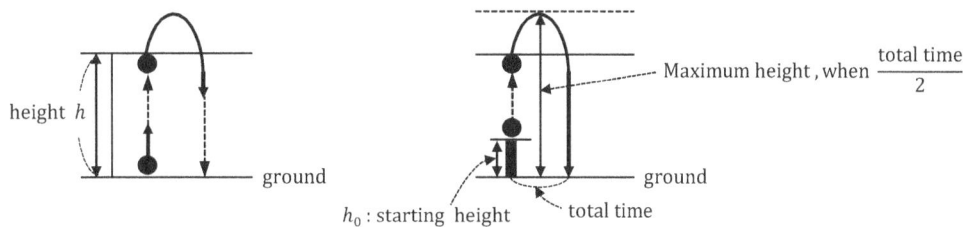

$60t - 5t^2 = 100$; $12t - t^2 = 20$; $t^2 - 12t + 20 = 0$; $(t - 2)(t - 10) = 0$

So, $t = 2$ seconds (upward) and $t = 10$ seconds (downward)

(2) When will the ball reach the ground again?

$60t - 5t^2 = 0$; $5t(t - 12) = 0$; $t = 0$ (depart) or $t = 12$ (arrive)

Since $t > 0$, $t = 12$ seconds

(3) What height will the ball reach in 4 seconds?

$h = 60 \cdot 4 - 5 \cdot 4^2 = 240 - 80 = 160$ feet

(4) What is the maximum height the ball will reach?

$\frac{0+12}{2} = 6$ seconds $\quad \therefore h = 60 \cdot 6 - 5 \cdot 6^2 = 360 - 180 = 180$ feet.

Solutions for Chapter 8

#1. Simplify each expression.

(1) $\dfrac{2x^5}{x^2} = 2x^{5-2} = 2x^3$

(2) $\dfrac{12x^4y}{56x^2y^6} = \dfrac{2^2 \cdot 3 \cdot x^4 \cdot y}{2^3 \cdot 7 \cdot x^2 \cdot y^6} = \dfrac{3 \cdot x^2}{2 \cdot 7 \cdot y^5} = \dfrac{3x^2}{14y^5}$, $x \neq 0$

(3) $\dfrac{8-2x}{3x-12} = \dfrac{2(4-x)}{3(x-4)} = \dfrac{-2(x-4)}{3(x-4)} = \dfrac{-2}{3}$, $x \neq 4$

(4) $\dfrac{x^2+6x+8}{x^2+3x-4} = \dfrac{(x+4)(x+2)}{(x-1)(x+4)} = \dfrac{x+2}{x-1}$, $x \neq 4$

(5) $\dfrac{3}{x} + \dfrac{1}{2x} = \dfrac{3 \cdot 2}{2x} + \dfrac{1}{2x} = \dfrac{3 \cdot 2 + 1}{2x} = \dfrac{7}{2x}$

(6) $\dfrac{2x}{x^2-4} - \dfrac{3}{x+2} = \dfrac{2x}{(x+2)(x-2)} - \dfrac{3(x-2)}{(x+2)(x-2)} = \dfrac{2x-3(x-2)}{(x+2)(x-2)} = \dfrac{-x+6}{(x+2)(x-2)}$

(7) $\dfrac{x+1}{x^2-9} - \dfrac{x}{x^2-2x-15} = \dfrac{x+1}{(x+3)(x-3)} - \dfrac{x}{(x-5)(x+3)} = \dfrac{(x+1)(x-5)-x(x-3)}{(x+3)(x-3)(x-5)} = \dfrac{x^2-4x-5-x^2+3x}{(x+3)(x-3)(x-5)}$

$$= -\dfrac{x+5}{(x+3)(x-3)(x-5)}$$

(8) $\dfrac{x+3}{x^2-4} \cdot \dfrac{x+2}{x^2+2x-3} = \dfrac{x+3}{(x+2)(x-2)} \cdot \dfrac{x+2}{(x+3)(x-1)} = \dfrac{1}{(x-2)(x-1)}$, $x \neq -3$, $x \neq -2$

(9) $\dfrac{x^2-16}{(x+5)^2} \div \dfrac{x+4}{x^2+2x-15} = \dfrac{x^2-16}{(x+5)^2} \cdot \dfrac{x^2+2x-15}{x+4} = \dfrac{(x+4)(x-4)}{(x+5)^2} \cdot \dfrac{(x+5)(x-3)}{x+4} = \dfrac{(x-4)(x-3)}{(x+5)}$, $x \neq -4$

(10) $\dfrac{\frac{2}{x^3}}{\frac{8}{x^5}} = \dfrac{2}{x^3} \div \dfrac{8}{x^5} = \dfrac{2}{x^3} \cdot \dfrac{x^5}{8} = \dfrac{2x^{5-3}}{8} = \dfrac{x^2}{4}$

(11) $\dfrac{\frac{2x^2-5x-3}{x-3}}{\frac{2x^2-7x-4}{2x}} = \dfrac{2x^2-5x-3}{x-3} \div \dfrac{2x^2-7x-4}{2x} = \dfrac{2x^2-5x-3}{x-3} \cdot \dfrac{2x}{2x^2-7x-4} = \dfrac{(2x+1)(x-3)}{x-3} \cdot \dfrac{2x}{(2x+1)(x-4)}$

$$= \dfrac{2x}{x-4} \text{ , } x \neq 3 \text{ , } x \neq -\dfrac{1}{2}$$

#2. Divide the following long rational expressions

(1) $(x^2 + 5x - 6) \div (x - 2)$

$$\Rightarrow \quad x - 2 \,{\overline{\smash{\big)}\,x^2 + 5x - 6}} \atop \displaystyle x+7$$

$$\begin{array}{r} x+7 \\ x-2\,)\overline{x^2+5x-6} \\ \underline{x^2-2x} \\ 7x-6 \\ \underline{7x-14} \\ 8 \end{array}$$

$$\therefore \quad \frac{x^2+5x-6}{x-2} = x + 7 + \frac{8}{x-2}$$

(2) $(2x^2 - 20) \div (x + 3)$

$$\begin{array}{r} 2x \quad -6 \\ x+3\,)\overline{2x^2+0\cdot x-20} \\ \underline{2x^2+6x} \\ -6x-20 \\ \underline{-6x-18} \\ -2 \end{array}$$

$$\therefore \quad \frac{2x^2-20}{x+3} = 2x - 6 - \frac{2}{x+3}$$

(3) $(3x^3 + 2x^2 + 1) \div (x - 2)$

$$\begin{array}{r} 3x^2+8x+16 \\ x-2\,)\overline{3x^3+2x^2+0\cdot x+1} \\ \underline{3x^3-6x^2} \\ 8x^2+0\cdot x+1 \\ \underline{8x^2-16x} \\ 16x+1 \\ \underline{16x-32} \\ 33 \end{array}$$

$$\therefore \quad \frac{3x^3+2x^2+1}{x-2} = 3x^2 + 8x + 16 + \frac{33}{x-2}$$

#3. Solve the following rational equations

(1) $\frac{3x}{2} = \frac{5}{8}$; $3x \cdot 8 = 2 \cdot 5$; $x = \frac{10}{24}$ $\therefore x = \frac{5}{12}$

(2) $\frac{2}{3x-4} = \frac{-1}{2-4x}$; $2 \cdot (2-4x) = -1 \cdot (3x-4)$; $4 - 8x = -3x + 4$; $5x = 0$ $\therefore x = 0$

(3) $7 + \frac{4}{x-1} = -2x$; $\frac{7(x-1)}{x-1} + \frac{4}{x-1} - 2x$; $\frac{7(x-1)+4}{x-1} = -2x$; $7x - 3 = -2x(x-1)$

$2x^2 + 5x - 3 = 0$

$(x+3)(2x-1) = 0$ $\therefore x = -3$ or $x = \frac{1}{2}$

(4) $\frac{1}{x+2} + \frac{2}{x^2-4} = \frac{4}{x-2}$; $\frac{x-2}{(x+2)(x-2)} + \frac{2}{(x+2)(x-2)} = \frac{4(x+2)}{(x+2)(x-2)}$; $\frac{x-2+2}{(x+2)(x-2)} = \frac{4(x+2)}{(x+2)(x-2)}$

$x = 4(x+2)$; $3x = -8$ $\therefore x = -\frac{8}{3}$

#4. Solve the following inequalities

(1) $\frac{x-4}{x+3} > 0$

Since ($x + 3 = 0 \Rightarrow x = -3$) and ($x - 4 = 0 \Rightarrow x = 4$),

the critical points are $x = -3$ and $x = 4$.

Case 1. $x < -3$

$\Rightarrow$ Choose a point $x = -4$. Then, $\frac{x-4}{x+3} = \frac{-4-4}{-4+3} = \frac{-8}{-1} = 8$

Since $8 > 0$, $\frac{x-4}{x+3} > 0$ is true.

Case 2. $-3 < x < 4$

$\Rightarrow$ Choose a point $x = 0$. Then, $\frac{x-4}{x+3} = \frac{0-4}{0+3} = \frac{-4}{3}$

Since $\frac{-4}{3} < 0$, $\frac{x-4}{x+3} > 0$ is false.

Case 3. $x > 4$

$\Rightarrow$ Choose a point $x = 5$. Then, $\frac{x-4}{x+3} = \frac{5-4}{5+3} = \frac{1}{8}$

Since $\frac{1}{8} > 0$, $\frac{x-4}{x+3} > 0$ is true.

Therefore, the solution is $x < -3$ or $x > 4$.

(2) $\dfrac{x+5}{x-2} \leq 0$

Since $(x-2=0 \Rightarrow x=2)$ and $(x+5=0 \Rightarrow x=-5)$,

the critical points are $x=2$ and $x=-5$.

Case 1. $x \leq -5$

$\Rightarrow$ Choose a point $x=-7$. Then, $\dfrac{x+5}{x-2}=\dfrac{-7+5}{-7-2}=\dfrac{-2}{-9}=\dfrac{2}{9}$

Since $\dfrac{2}{9}>0$, $\dfrac{x+5}{x-2} \leq 0$ is false.

Case 2. $-5 \leq x < 2$ ($\because x \neq 2$)

$\Rightarrow$ Choose a point $x=0$. Then, $\dfrac{x+5}{x-2}=\dfrac{0+5}{0-2}=\dfrac{5}{-2}$

Since $\dfrac{5}{-2}<0$, $\dfrac{x+5}{x-2} \leq 0$ is true.

Case 3. $x>2$ ($\because x \neq 2$)

$\Rightarrow$ Choose a point $x=3$. Then, $\dfrac{x+5}{x-2}=\dfrac{3+5}{3-2}=\dfrac{8}{1}=8$

Since $8>0$, $\dfrac{x+5}{x-2} \leq 0$ is false.

Therefore, the solution is $-5 \leq x < 2$.

(3) $\frac{x}{2} \geq \frac{5}{x-3}$

$\frac{x}{2} - \frac{5}{x-3} \geq 0$; $\frac{x(x-3)-10}{2(x-3)} \geq 0$; $\frac{x^2-3x-10}{2(x-3)} \geq 0$; $\frac{(x-5)(x+2)}{2(x-3)} \geq 0$

Since $(x - 3 = 0 \Rightarrow x = 3)$, $(x - 5 = 0 \Rightarrow x = 5)$, and $(x + 2 = 0 \Rightarrow x = -2)$,

the critical points are $x = 3$, $x = 5$, and $x = -2$.

Case 1. $x \leq -2$

$\Rightarrow$ Choose a point $x = -3$. Then, $\frac{x}{2} = \frac{-3}{2} = \frac{-9}{6}$ and $\frac{5}{x-3} = \frac{5}{-3-3} = \frac{5}{-6}$

Since $-\frac{9}{6} < -\frac{5}{6}$, $\frac{x}{2} \geq \frac{5}{x-3}$ is false.

Case 2. $-2 \leq x < 3$ ($\because x \neq 3$)

$\Rightarrow$ Choose a point $x = 1$. Then, $\frac{x}{2} = \frac{1}{2}$ and $\frac{5}{x-3} = \frac{5}{1-3} = \frac{5}{-2}$

Since $\frac{1}{2} > \frac{5}{-2}$, $\frac{x}{2} \geq \frac{5}{x-3}$ is true.

Case 3. $3 < x \leq 5$ ($\because x \neq 3$)

$\Rightarrow$ Choose a point $x = 4$. Then, $\frac{x}{2} = \frac{4}{2} = 2$ and $\frac{5}{x-3} = \frac{5}{4-3} = 5$

Since $2 < 5$, $\frac{x}{2} \geq \frac{5}{x-3}$ is false.

Case 4. $x \geq 5$

$\Rightarrow$ Choose a point $x = 6$. Then, $\frac{x}{2} = \frac{6}{2} = 3$ and $\frac{5}{x-3} = \frac{5}{6-3} = \frac{5}{3}$

Since $3 > \frac{5}{3}$, $\frac{x}{2} \geq \frac{5}{x-3}$ is true.

Therefore, the solution is $-2 \leq x < 3$ or $x \geq 5$.

(4) $\dfrac{2x^2-5x-3}{x-4} > 0$

$\dfrac{2x^2-5x-3}{x-4} = \dfrac{(2x+1)(x-3)}{x-4} > 0$

Since $(x-4=0 \Rightarrow x=4)$, $(2x+1=0 \Rightarrow x=-\frac{1}{2})$, and $(x-3=0 \Rightarrow x=3)$,

the critical points are $x=4$, $x=-\dfrac{1}{2}$, and $x=3$.

Case 1. $x < -\dfrac{1}{2}$

$\Rightarrow$ Choose a point $x=-1$. Then, $\dfrac{2x^2-5x-3}{x-4} = \dfrac{2+5-3}{-5} = -\dfrac{4}{5}$

Since $-\dfrac{4}{5} < 0$, $\dfrac{2x^2-5x-3}{x-4} > 0$ is false.

Case 2. $-\dfrac{1}{2} < x < 3$

$\Rightarrow$ Choose a point $x=0$. Then, $\dfrac{2x^2-5x-3}{x-4} = \dfrac{-3}{-4} = \dfrac{3}{4}$

Since $\dfrac{3}{4} > 0$, $\dfrac{2x^2-5x-3}{x-4} > 0$ is true.

Case 3. $3 < x < 4$

$\Rightarrow$ Choose a point $x=3\dfrac{1}{2}=\dfrac{7}{2}$. Then, $\dfrac{2x^2-5x-3}{x-4} = \dfrac{\frac{49-35-6}{2}}{\frac{7-8}{2}} = \dfrac{\frac{8}{2}}{\frac{-1}{2}} = -\dfrac{16}{2} = -8$

Since $-8 < 0$, $\dfrac{2x^2-5x-3}{x-4} > 0$ is false.

Case 4. $x > 4$

$\Rightarrow$ Choose a point $x=5$ Then, $\dfrac{2x^2-5x-3}{x-4} = \dfrac{50-25-3}{5-4} = 22$

Since $22 > 0$, $\dfrac{2x^2-5x-3}{x-4} > 0$ is true.

Therefore, the solution is $-\dfrac{1}{2} < x < 3$ or $x > 4$.

Index

A

Absolute values, 19

Adding and subtracting polynomials, 53

Adding and subtracting rational expressions, 160

Algebraic functions, 158

B

Balance, 17

Base, 50

C

Closed points, 35

Coefficient, 16

Combined inequalities, 36

Complex fraction, 163

Compound inequalities, 36

Constant, 16

Critical numbers, 166

Cross product (multiplication), 17, 165

D

Denominator, 160

Discriminant D, 141

Distributive property of exponent, 51

Dividend, 162

Dividing, 53, 55, 161

Dividing rational expressions, 161

Divisor, 162

Double root, 136

E

Elimination method, 79

Equation, 16

Expanding exponents, 52

Exponents, 50

Expressions, 16

F

Factorization, 116, 121

Factorization formulas, 117

Formulas, 56

G

Graphs of linear inequalities, 102

Graphing systems of inequalities, 104

Graphing a system of equations, 86

Greatest common factor (GCF), 116

H

Half-plane, 102

I

Inequalities, 32

L

Leading coefficient, 118, 119

Least common denominator (LCD), 160, 165

Linear equation, 18, 20, 78

Linear inequalities, 34, 37, 100, 102

Long division of polynomials, 162